HOW LONG DID THE PONIES LIVE?

The Story of the Colliery at
KILLINGWORTH
and
WEST MOOR

by
Roy Thompson

Published by:
Beacon House
42 Beverley Terrace
North Shields
NE30 4NU

Typeset and Printed by
Craig Typesetting, Artwork & Print
321J Mayoral Way, Team Valley, Gateshead
NE11 0RT

This book is dedicated to Jack and James Thompson
of Australia, so that they might know how some
of their forefathers lived

ISBN 0 9530600 0 4

INTRODUCTION

Killingworth Colliery was for years, and perhaps still might be, the world's most famous mine.

In simple terms of being heard of by the most people during the last Century, or the present, it must rank supreme in any list of well-known centres of mineral extraction, which could include the Rand, the Klondyke, Rio Tinto, Mounts Morgan, Isa and Lyell and the fabled mines of King Solomon! For the best part of eighty years, after sinking was started in 1802, it produced coal, not in large quantities by modern standards, but sufficient to become famous on the London Market as the best of household coals - "Killingworth Wallsend" - and for the first half of its life led the technology of the mining world, thanks to the fortuitous combination of resident early locomotion pioneers, and Nicholas Wood, the leading Mining Engineer of the mid Nineteenth Century. But it is for being the home of George Stephenson that Killingworth Colliery is so much mentioned in books on industrial history, a lot being made of his still-standing humble cottage with his sun-dial, and the railway running past it of a gauge chosen by him of 4´8½″ for the colliery waggons and now standard worldwide.(1)

The pit itself has disappeared, the exact positions of its West Moor shaft unknown even to NCB surveyors in 1962 to the author's knowledge. A modern industrial estate stands in its place, and all trace of its rows of miners' houses has gone forever, as has its once extensive spoil-heap.

The Local Council has made a park with sign-boards, and mounted a pulley wheel at the High Pit near Killingworth Village, originally only a side venture of the first company and which had a separate life of its own for a while, and which ended up as the up-cast shaft for the Colliery, whose main surface appurtenances were a mile away at West Moor, adjacent to the Colliery Rows.

Killingworth, or West Moor Colliery as it came to be called at the end of its days, was one of the pits sending coal down waggonways to the Tyne

5

to be shipped to London.

Along with Jarrow, Hebburn, Felling, Wallsend and Heaton it was one of the "slaughter-pits" of the Tyne, so-called by Ellen Wilkinson, the Labour M.P. for Jarrow, when she wrote a history of that town in the 1940's, "The Town that was Murdered".

Many of the former inhabitants of Long Row, Short Row, Lane Row, Cross Row, Crank Row, High Row, Paradise Row and Office Row, formerly Quality Row, still survive, of course, but nobody now can remember the pit working. Burradon, originally one of Killingworth Collieries, which survived until 1975, and was worked long before the date of 1837 published as its starting date by the National Coal Board when it closed, carried on the technology, lore, vocabulary and even accent of the Killingworth Collieries as a living record, much of the workforce having moved from West Moor. Compared with mines further north in the County, sunk later in the Nineteenth Century, there was a notably old-fashioned social atmosphere at Burradon, when the author worked there during the 1950s, not that the working methods were unusually antiquated for the times (although the coal was still hand-filled and piles of caunch stones were shifted by men on their knees shovelling one to the other!)

As children maintain a folklore of games and rhymes, and even vocal tones unknown or forgotten to their parents, so Pits carry on the terminology and tales of bygone years, and use names and pronunciations for their tools and trades, long ago become obsolete and unmentioned at the surface. As a standard weight or measure is kept in a glass cabinet for ultimate reference, whatever the vagaries of the market place, so, provided there are not too many outsiders entering it, does a pit maintain the accent of a village. Thus was the harsh Geordie accent of West Moor, which had been commented upon by Victorian chroniclers of the Stephensons and by Newcastle newspapers of the day, preserved like a living sound archive at Burradon, a small colliery of inter-related families mostly descended from the Nineteenth Century workmen of Killingworth Colliery.

The aim of this little book is to collect the available written information, avoiding repetition of the well-known facts about George Stephenson, Killingworth's most famous inhabitant, and to present it without too much reference to technical detail which might be difficult or tedious to the general reader. Hearsay tales have been avoided, except where they could be cross-checked for probability of fact.

Some of the information here presented will seem commonplace to retired mining men, but some, to younger readers, may be unbelievable. Nevertheless it is all true. Today there is very little heavy labour, even in mines, of the sort described in these pages: workmen are no longer trained athletes, kept to a peak of physical toughness by their jobs, used to exercising skill with their muscles as they pull, push, lift and swing. Machines and the economy of scale have taken over, and such coal mines as remain are large, bustling places, the ammoniacal smell of horses, the silence, broken only by the chipping of a pick and the scrape of a shovel, replaced by the violent noise of conveyors, loud signalling systems and multi-horsepowered coal-getting machines. Away from the pit, a miner can hardly now be recognised, the old cloth-capped, hollow-jawed leanness, bandied legs and sand-paper hands replaced by well-dressed burliness, and only perhaps a touch of black coal dust in the corner of the eye, left by a hasty pit-head hot shower, betrays his calling; his son, if he has a job at all, might work in a supermarket, where, when he has to move a bogey loaded with beer cans, has difficulty because he does not know how to put his head down between his arms and push with his legs as did his putter forebears.

The pits have gone and memories are fading. An often garbled folklore remains, centred on the Pre-War years, of hardship and unemployment, ready material for reinforcing the socialism of the former mining districts, dwelling on tin baths, proggy mats, poss-tubs, poverty and short commons, in other words dealing with the world of the Colliery Row and home and family.

I hope here to cover the working life of the miner himself, and in earlier days; the miner was never prosperous, but in the context of the times he fared not too badly, earning more than any other semi-skilled worker of the Nineteenth Century. When Killingworth Colliery closed, in 1882, the mining population in Britain was growing rapidly with the Industry, which was to reach its all-time output peak in 1913; there were more men miners by that time than in any other occupation.

An over supply of cheap labour delayed mechanisation; the complacent attitude of the mining capitalists was indulged by the readiness of the miners, isolated in their pit villages from other workers, to cope with working conditions, and a degree of physical labour, even then unusual. An industry, whose technology had led the world a hundred years before, began to stagnate; productivity per man employed had dropped since the 1880's and after 1913 the number of men fell away. Nationalisation could not stop the decline, nor could mechanisation, when it came, make more than a few pits economic. Safer and better conditions of work and higher pay for the miners simply closed more mines; the coal could not pay for that.

As a young man I worked in the mines as a trainee mine-manager and had a glimpse of a rapidly disappearing world which seemed at the time to have been fossilised in 1914. There were mines, one knew, in "the Midlands" that were modern and mechanised, but by the early days of Nationalisation, in the soon to be closed pits of the North East, nothing much had happened, apart from electrification and mechanical coal-cutting, since the First World War. The miners (and the managers) had fought in that war and after it were glad of jobs which no doubt seemed very desirable after their experience in the trenches. To me it seemed that, apart from mining being much safer than fighting, there were many similarities to warfare: the line of command of management structure; the supplies system; the "kitting-up" with lamps, piecework tokens, drilling-bits, caddies of explosives, tins of detonators and other light supplies before descending the mine; the tramp inbye of large numbers of men together, heads down or even bent at the waist because of the low roof, often for several miles, some leading horses, carrying drinking water, tools, surveying instruments, occasionally each calling out to the man behind to warn of some obstruction such as a fallen stone or a loose wire-rope; the neat uniform of the officials, many with expensively tailor-made pit-suits carrying yard-sticks and the uniformity of the men in their cloth-caps and "hoggers" (the workforce were always referred to by the management as "the Men"); the thinning out of the ranks as datallers dropped off to man their haulage or supplies posts; and finally the arrival at the coal where facemen stripped off to crawl along the thin seam, caring for their lives as they went. When I tried to point out this similarity to an old soldier of the First War I nearly had my head bitten off for my impertinence to compare the two, he, no doubt, having memories of great danger and fallen comrades. But, safety apart, the analogy was definitely there, and I also think the presence of ex-officers in Colliery management between the Wars may have contributed to the perpetuation of tough working conditions underground; it certainly led to much of the terminology employed and the attitudes prevalent. One obvious manifestation of this toleration of hardship was the very low height of travelling roadways, kept so to save money. This was to me a continuing source of fury but was apparently ignored by my inured companions, one of whom to my early-days queried complaint replied, "When you're marching you aren't fighting!" Of course, in Roman days, miners at Rio Tinto in Spain drove tunnels two and a half feet high and narrow enough to prevent a man turning around to avoid crawling out backwards. (An entombed chained skeleton was found by the British Company who reworked the pyrites in the Nineteenth Century.) Things were always worse in the past!(2)

This is the story of Killingworth Colliery, which closed when Queen Victoria had still eighteen years to reign and mining seemed a permanent way of life.

FOOTNOTES

(1) This was reputedly taken from old Roman chariot-wheel marks, left in stone at the nearby Hadrian's Wall, but this story may be apocryphal; it is obvious that the gauge chosen would have to match the existing horse-drawn railway to the Tyne.

(2) The worst of these low travelling ways was that of the Beaumont Seam, Rising Sun Colliery: two miles at tub height.

Acknowledgements

My thanks are due to the following persons and organisations:

North of England Institute of Mining and Mechanical Engineers

North Tyneside Library, North Shields

Newcastle Central Library

Northumberland Record Office

Dennis Dixon

Norman Whitelock

Maurice Dobson

J.R. Crone

The Ordnance Survey

The Coal Authority

Tyne and Wear Archives Service

Gordon Swinton Photographers

Peter Martin

Aidan Doyle

Craig Typesetting

Michael Anderson

Maggie McGregor

Claire Victoria Robertson

D. Clarke

Eric Hollerton

West Moor Pit Head C. 1850
*An old engraving, somewhat out of perspective, but
showing two large pulley wheels for the steam
winder and another two, smaller, for the horse-gin.
The window in the winding-engine house is
identical to those in the locomotive shed still
standing in the 1940's and the picture may be
"composed".*

[North Tyneside Libraries]

CHAPTER 1 A Georgian Mine

In 1802, Killingworth Moor was the name for the area of agricultural land lying to the north of Benton and Gosforth, which already had their own profitable collieries, rising very gently and evenly away from the Tyne Valley, across the Great Lime Road, which marked the limit of the Parish of Longbenton, and it covered about five or six square miles. It stretched from the woods of Gosforth Park, the home of the Brandling family, coal owners, eastwards to Scaffold Hill where a stand was at one time erected for the watchers of the Newcastle Races, before they moved to Blaydon after the Moor was enclosed in 1790. Trial boreholes had been put down in several places there during the previous century and it was known that the High Main Seam, the then equivalent of today's North Sea Oil lay below. The Grand Allies, as they were known, led by Sir Thomas Henry Liddell (Lord Ravensworth)(1), ventured some risk capital and began the sinking of Killingworth Colliery, well away from the already working Prosperous Pit near Scaffold Hill, at the West Moor, not far from the agricultural Killingworth Village. The new shaft, which was to have the latest in pumping and winding engines, (the Prosperous Pit drew coals by means of a water wheel powered by water pumped from the mine to above its level!) was referred to in the plural: the A and B shafts. A wooden brattice divided the shaft throughout its length to provide a two-way passage of ventilating air; the steam driven winder could draw corves of coal up one side, the A pit or downcast and the B Pit, or upcast would house the rising main water pipe powered by a string of connected rods successively lifted and dropped at the end of a huge beam whose other end was connected to a large piston in a steam cylinder. This "B pit" eventually would also house insulated steam pipes to supply power to underground machinery.

On July 12th, 1805 the "engine" shaft cut the High Main Seam at 637 ft. As the sinkers dug down into the seam, which they knew they had reached by the well-known roof strata they were passing through, they found first, 4 inches of dull hard, top or cannel coal, then 6 ft. 1 ins. of bright household coal and then 8 ins. of bottom coal, softer than the rest but still worth working for its easier hewing qualities, although a lot of it might have to be discarded as smalls, or used for the colliery engines.

A "thill" or "band" came next, which could either serve as a floor for working the seam or be removed and stored, giving access to a following 3 ft. of coal. It was the richest seam of coal ever encountered in this country (2); there might be thicker seams in Staffordshire or elsewhere, but the High Main of the Lower Tyne basin was the greatest concentration of easily-worked, supreme quality, household coal of all time, readily transportable to London by a network of wooden rails bearing chaldron waggons which could mostly gravitate to the river, where the coal was transferred to sailing brigs via 21-ton keel boats.

By the time the first shaft at Killingworth was sunk, the High Main was nearly finished at some of the riverside collieries, where wasteful methods of working had sometimes been used. Those owners with unworked areas of the seam were envied. Further north than Killingworth it thinned down before outcropping, and south of the river, in Durham, it developed a band of barren rock which increased in thickness so that it could not be worked as one seam. But on the West side of Killingworth Moor in 1802 the High Main, which came to be sold in London as "Killingworth Wallsend", was lying intact in all its virgin glory, ready to be raped and pillaged. During the first sixty years of the Colliery's life no other seam was worked. There would be some steep gradients but these could be readily overcome (and were) by a plentiful supply of cheap labour.

The shaft, which had a finished diameter of 14 ft, large for those days, was continued down to 720 ft, partly by way of exploration, but principally to provide standage for water prior to pumping it to the surface. Where it was considered necessary the shaft was lined with stone walling, but some of the hard "post" or sandstone would be left to support itself. Any wet strata passed through would be sealed with wood planks held in position by oak cribbing or possibly by the new iron rings which were beginning to be used at some pits. Gunpowder would be used sparingly, because of the difficulties and delays in igniting it by a glowing red-hot iron washer sent down a string held taut from the surface, most of the excavation being by

14

heavy picks, wielded by cheap labour, rhythmically chipping away the stone to the circular profile required. The bigger pieces of sandstone from the shaft would be used to help build the first cottages, to house the sinkers, and then the men and families would follow to work the coal.

Later the shaft was deepened to1044 ft. to below the Low Main Seam.During its sinking the shaft would be divided into two by a brattice made of wood, to provide air for the sinkers. Shortly after it was finished, however, two more partitions were added. This is known from George Stephenson's description of an explosion in 1806. One of the partitions would be for the Engine. This means the pumping engine, the reason the shaft was called the Engine Pit. The engine would be built on brick foundations at the surface and was a low-pressure beam engine, the atmosphere forcing a large piston into an evacuated cylinder. The beam lifted on each stroke a string of rods or "spears" which pumped the mine water up a cast iron pipe running the length of the engine compartment. Probably at that time two of the other compartments were used for winding coal, a rope down each so that each set of corves was separated, empties from fulls, so that they did not meet mid-shaft. This would also half the time needed for single winding. The weight which could be suspended on each rope would be limited and this method could also reduce the load on the winding engine.

However, we know that this system did not last long and old prints show both the Engine Pit and the later High Pit, each equipped with an engine counterpoised with a rope leading to a heavy chain in a separate shaft or well. Given stronger ropes a heavier load could be lifted and the weight could be balanced in this manner, putting less strain on the engine. The chain on the counterpoise rope was arranged so that the links could pile up at the bottom of the well, so that a descending corf had just to lift the rope and then the links, one by one, from the well and by the time all the links were freely suspended it had accelerated to speed. In the case of an ascending full corf the engine had the assistance of the chain from the start, but when the corf reached the surface it lost it gradually and the extra strain on the engine would help in braking. A bell was fitted to warn of an overwind.

It is impossible now to be sure when and how the systems were changed but it is likely that Stephenson had a say in matters. We know that an early engine was lying at Killingworth in the 1880's, the cylinder of which was marked as having been cast at Coalbrookdale in 1767. Possibly after the explosion at the shaft bottom in 1806, the winding arrangements had to be rebuilt. Nicholas Wood's student notebook of 1812 refers to the A and B pits at West Moor and the C and D pits at Killingworth Village i.e. "two pits" per shaft. We know from a later paper, written in 1855 by Wood, that a beam engine had been used in the early days for drawing coals and it is probable that it was superseded early in the life of the Colliery by a double-acting high-pressure engine, following James Watt's principles. (Watt's patents had expired just when the mine was started. A tall engine tower, typical of such an engine, survived the eventual closure of the mine.) Although the headgear at West Moor was also fitted with a horse-gin and pulleys for much of the life of the shaft, only steam was ever used for drawing coal, and the horse-gin was only for shaft-fitting and supplies. The main headgear pulleys can be seen in an old print, above those for the horse-gin and are in line with the pulleys over the counterpoise well.Towards the end of its life, certainly by 1862, the shaft wound four cages, with two steam winders: from the headframe bracing it can seen that the second, smaller winder was in front of the first and slightly to one side.

However, the very first winding engine was a beam engine with a crank to produce rotary motion. It must have been situated away from the vertical Watt-type engine, on the other side of the shaft: the crank gave its name to Crank Row, the central row of houses at West Moor Village.

The fourth compartment was for upcast ventilation, a furnace built at the bottom created a convection air-current which could draw fresh air down the other compartments. At a later stage, when there were only two compartments, the shaft had furnaces at two levels. Much later, the entire shaft became the downcast-air and winding shaft, the upcast-air coming out at the High Pit at Killingworth Village, where a furnace was kept burning.

After 1812, when the West Moor shaft had two divisions, the upward air current would be diverted through a drift to a shallow, second shaft, above which a brick chimney was built. This kept the furnace fumes away from work at the point where coals were landed and also prevented recirculation of foul air.

The first Burradon Shaft was completed in 1819, when George Stephenson was involved in building a pumping engine there, and was producing coal by 1820. It is likely that there was soon a connection underground, but, so far as is known, the ventilation was separate.

By midsummer 1805 the winning was completed, coal began to come out of the shaft, the Earl of Carlisle began to receive his royalties and the village of West Moor began to grow, firstly as a long row along the Great Lime Road, where it was eventually to be split by the main line railway. Miners came from all parts as production increased, at first from neighbouring collieries and eventually from afar: starving Irishmen; Cornish tin-miners when strike-breakers were needed; cholera-driven, lantern-jawed agricultural labourers from the south, skeleton-limbed Scots escaping from the feudal semi-slavery of the northern mines; families turned out of their tied cottages by Northumbrian farmers. They were replaced often, over the years, by incoming families, but eventually a much inter-related community stabilised, to remain largely together over the next century.

The coal was worked leaving pillars. It was worked whole: it was too high to safely hew underneath more than a foot or two, but still huge slabs could be levered down, to be easily broken up to fill wicker baskets which were dragged on sledges to the shaft bottom, to be tipped into corves (q.v.). Later the corves were mounted on trolleys and filled directly at the coal-face.

The temptation for the management was to leave small pillars and this resulted in the soft rock below the seam heaving, rising like solid porridge to fill the working place within a few days, making it inaccessible. New

17

Killingworth High Pit
This engraving must date from the 1840's.
The building to the left would be the original pumping
engine, now disused. The headgear to the
extreme right is the counterpoise.
The horse-gin would be for shaft work and supplies.
The chimney in the distance could be
the Prosperous Pit. The hills in the background were
Co. Durham, across the River Tyne, mainly artist's
licence, but the woman carrying water would
be real enough.

[North Tyneside Libraries]

places would then be started to keep up output. The result was that good mineable coal was lost. In the 1860's a new method of working was used to rework these old pillars. This heaving of the floor or "creep" (q.v.) was not the only hazard to output. From the start Killingworth was a gassy pit.

1806

George Stephenson was present to see the 1806 explosion. By that date he was the winding engineman. He described what happened thirty years after the event to a Parliamentary Commission. The pit had just stopped drawing coals for the day and four men had descended to work on the shaft-bottom furnace. Stephenson had just lowered another man who was to supervise the work when, within three minutes, there was an explosive discharge as though the shaft were a cannon. For a quarter of an hour wood-brattice and stones shot from the shaft, and trusses of hay went up into the air like balloons. Clearly the brattice which divided the fourteen-feet shaft into four quadrants was blown away in part. When the firing stopped, the surrounding surface atmosphere was then drawn in a rush to fill the partial vacuum left in the mine. Stephenson then lowered the rope repeatedly down, pausing each time, to allow anyone alive to grab a hold. Miraculously, several men got out, one landing without injury after being blown up the shaft on the slackened rope by another explosion. In all, ten men died in the mine including four of the five men he had lowered; their bodies were not found until twenty-three weeks later. The blasts continued for two days, every two or three hours sufficient fresh air being sucked into the mine to enable the firedamp to re-explode. The loss to the owners was, he said, £20,000. Stephenson's description tallies with that of John Buddle, who witnessed an explosion at Harraton in 1817 and described a black cloud like a water spout rising from the shaft for about five minutes.

1809

Another explosion occurred, this time inbye, and twelve men were killed. Meanwhile the village grew. At the beginning the houses were single-storey with earth floors. There were no privies, only dung-heaps between the rows, eventually cleared by farmers to fertilise their fields. Nor would there be any running water from pipes, although it is likely that hot water would be available from the Colliery Yard as effluent from the steam engine. Water would be collected from ponds or into a barrel from the roof, although at least one well was soon provided. Water had always been available for the agricultural community of Killingworth Village, where there were three farms, by a tiring uphill carry from the Letch Well, a pool where a stream crossed the Great Lime Road, the direct path nowadays still maintained as a right of way. Later, under the benevolent management of Nicholas Wood, open drainage channels, flushed by water from the mine or rainwater, would follow the line of each row of houses. Water was never delivered to each house, nor did the houses ever have foul-drains during the life of the mine. The roofs of the houses were red pantiles of the traditional Northumbrian design. Only after the 1870's were these tiles replaced by grey Welsh slates, as extra storeys and lean-tos were added. There was no paving and only gradually was the heaving quagmire, which developed after rain, turned into a hard black surface of small coal, unsaleable and otherwise unusable.

But those improvements lay in the future. In 1805, West Moor Village was an open working place with no pretensions to nicety. Work was offered on a take-it-or-leave-it basis, and families, driven by desperate poverty and even famine elsewhere, rushed in to take it. It is difficult for us now to imagine what life there must have been like at that time. Certainly there must have been a pioneering spirit, with people glad to have arrived and settled and no longer hungry. But it was no gold-rush with hopes of riches: only an escape from grinding hunger. There was not even the escape from the humiliating yearly hiring burden of agricultural labour: the aspiring miner signed a yearly bond with the mining company, which re-entered him into a strict servitude by which he could be jailed should he transgress its harsh provisions. At least it was only for a year whereas in Scotland, where serfdom still prevailed, the miner was bound for life and could be sold with the colliery to a new owner. But the miner worked alone, largely out of sight of his masters and left to his own devices, paid by results and not begging and scraping to some small farmer. And he was well paid, comparatively. Wages at the beginning of the Nineteenth Century were often actually higher than those of late Victorian times, for coal-owners

20

would compete for labour and even offer down-payments for the signing of the Bond when coal was fetching good prices on the London market. He had a house provided and he was warm! His fire never went out, he had an endless supply of free, small coal. His sons went to work in the mine at a very early age, although a school was soon sponsored by the Company. He dressed well, was even a dandy, but he might be coal black underneath and was even acceptable with a black face in his best suit, a surprisingly common sight in the very early 1800's!

Although his sons worked alongside him underground, his wife and daughter never did. Elsewhere in Britain this was not always the case but in Northumberland, during the whole of the Nineteenth Century, females were never employed, either underground or on the surface. Nevertheless, life was brutish and short: illness was prevalent, mortality was high, caused not least by accidents at work which were very numerous. We can only surmise that the miner of 1805 went to work with some trepidation. The very act of stepping into a loop on the end of a hempen rope, above the deep shaft, before being dropped at high speed would be frightening at first; all sorts of dangers awaited him below. Pitmen developed bravado and it is no wonder that they embraced religion as evangelism increased into the Nineteenth Century. In the very early days of West Moor it is likely that every penny available after necessities had been paid would go on drink. Order was kept by the Company's own guards, in the absence of an independent police force. Magistrates were all powerful and could call in the militia for any disturbances. It is known that Dragoons were stationed in Newcastle in the 1820's to cope with serious political dissent among the miners, as they became organised and caught a whiff of the spirit of revolution then abroad in Europe. No doubt the early mining families turned out to see the troops and listened to the rattle of sabres as the horses clopped along the Great Lime Road. At the beginning, the Napoleonic Wars were in full spate: the battle of Trafalgar was three months after the first coal was wrought; George Stephenson had invented his safety-lamp before Waterloo. (He narrowly avoided call-up for the Militia by paying for a substitute.) After the public house the main leisure activity was gardening. Not the ornamental kind but vital growing of vegetables. As

West Moor Colliery

1840's with some artistic licence. The counterpoise can be seen on the other side of the winding engine house. The engraver may have been confused as the rope to the shaft improbably comes out horizontally from a side door. Next to the main pulley is the ventilating chimney. To the right is another chimney to send steam down the shaft. The pulley wheel to the extreme right is the Crab or heavy-duty crane. The tall engine house, of boarded wood stood until after the mine closed.

[North Tyneside Libraries]

Methodism spread, and Temperance flourished, gardening assumed increasing importance in the life of the village and whole fields were laid out as allotments by the Company for the benefit of their most-favoured, longer-serving employees. However, beer was always an incentive for the miner to work harder and the ability to pay for a few pints per week was regarded as the difference between austerity and the good life.

The miners were never town-dwellers, and saw themselves as countrymen, and apart from their gardens they would supplement their diets when possible by gleaning and poaching, and worked for farmers when the pit was idle. The better-off kept a pig.

The miners had little social status. People from outside the mining districts did not regard them highly; the middle classes within, the doctors, clergymen, shopkeepers, farmers, found them quaint and amusing, the "salt of the Earth", but never really knew them. Looking now at verbatim records of inquests, one often admires the intelligence of the bullied witnesses and their occasional spirited repartee. The pitmen then, were not at the bottom of the heap because of any natural selection: unless they were gifted to a scale of genius, they had no chance of moving up the social ladder. Only free secondary education, when it came a hundred years later allowed the cleverer boys a chance to leave the mines. There must have been many clever men, with good brains, who might have been university professors, who laboured away day by day at back-breaking repetitive work. Because the men who worked together, lived together in small isolated villages, they grew alike, but together they were out of touch with the working class of the towns and factories and shipyards.

Some names persist at Killingworth from the earliest times: Carr, Hedley, Hepple, Hindmarsh are notable examples. At first the turnover in families was greater than happened by mid-century, when the industry was more settled. Many more names, still living at West Moor a hundred years later, then appeared, examples of which are Anderson, Dixon, Dryden and Mason. These are common enough names in the North East but Cooperthwaite and Trevelyan are surely indicative of ironstone- and tin-mining origins.

23

(Strike breakers were brought in from Cumberland and Cornwall as well as Ireland.) In the end they all blended and became Geordies, (George was indeed a very common Christian name at Killingworth) sons following fathers down the pit.

Mobility upwards may have been difficult but social descent was certainly easy in the Nineteenth Century. It always has been, and it is likely that we are each of us descended from some King of not too long ago. Because infant mortality has been higher, and reproduction lower, among poorer people since life began, the genes of the working class must have been steadily mixed over the centuries with those of their masters*. It is easy to see how a miner could be descended from a lord. Title and property always passed to the first-born, and the second son of a second son of a large landowner might be a tenant farmer, and then his second son might become, say, a contractor supplying corves to collieries or a master shaft-sinker and perhaps settle in a mining village. Then, if his business did not prosper, his sons, first, second or fifth were likely to start as putters and go on to become hewers. And their sons would consider themselves miners born and bred. There were plenty of jobs for them as the Coal Trade expanded. And miners they would stay, until modern times brought better education.

The isolation of miners from the rest of the world has always brought solidarity and cast-iron Unionism. They have had to rely on each other for their pay as well as for their safety; they are tough on transgressors. They are the same all over the world, they even look the same! A colleague during his National Service in Cyprus in the 1950's saw miners waiting for a bus outside a copper mine and thought for a moment he was back in his native Durham. In the Northern Territory of Australia the author recently saw, in the midst of old gold diggings, evidence of desperately hard work in the 1930's, a little monument raised by the people of the town to the memory of miners everywhere. They are a breed and their wives are part of the breed.

Miners' wives at West Moor led a life of unremitting hard work, which has been much chronicled elsewhere, so suffice to say that the inside of a

24

miner's cottage was always spick and span, the front step pipe-clayed, the fire-place black-leaded, furniture treasured and polished. Meals and baits were prepared whenever required; baking and washing done on set days. The wives made a point of getting on well with their neighbours: doors were always open, bread was baked together, washing "double-possed". A miner's mother would be shocked if he married a girl not a miner's daughter; only girls bred to the life could cope with it.

As in all villages, everyone knew everyone else. That is not to say that they always got on well together. Arguments happened, and were quickly settled by fights, which were common occurrences. Gambling was common but became illegal as time went on. Pitch and toss schools were vigilantly persecuted by the police in the second half of the Century. Bookmaking was carried out surreptitiously, by publicans or, more recently, at the village shop, which was largely sustained by betting. Later in the Century the Racecourse at Gosforth Park led to some non-mining employment: the large Killingworth Station delivered horses as well as race-goers and had to be staffed, as had the telegraph; retired miners manned the turnstiles on race-days. Mary Todd, widowed by a pit accident, avoided working in the fields to keep her family together by her home-dressmaking skills and sewing jockeys' silks.

*This process has now ceased or even reversed, in the author's opinion, due both to improvment in medical care for children and changed political outlook worldwide, permitting class intermarriage.

GEORGE STEPHENSON'S COTTAGE.

FOOTNOTES

(1) The two other main partners were Sir William Augustus Cunningham of Richmond, Surrey and the Rt. Hon. John Bowes, the Earl of Strathmore. Their original royalty agreement was with the main landowner, the Earl of Carlisle, guaranteeing a certain rent of £750 for 1000 tens of coals over a term of 63 years. Surplus loadings to that would incur charges of 15s per ten of 22 waggons, each of 19 bolls. A time for the winning, free of rent, for three years was agreed. Similar arrangements were made with other, smaller landowners, particularly a Mr. Punshon and a Mr. Bonner of Killingworth Village; where necessary "outstroke" or wayleave rents of 2s 6d per ten and a shaft rent (for the High Pit) of 2s 6d per ten were paid, these payments being allowed to make up "shorts" should the output fall below the agreed certain rent. In 1821 Mr. Bonner took the Allies to Court claiming that part of the High Pit output was through his land. From the Matthew Liddell letters it appears that a royalty of 27s 6d per increased ten of 440 bolls was payable in all to the Earl of Carlisle, the size of waggon being by then 20 bolls.

(2) A labelled block of High Main coal was kept on display in the Rising Sun Colliery, Wallsend, surveying office when the author was there in 1950, the seam itself having been, then, long since worked out. Every pit manager dreamed of getting into a pocket of the High Main, but at the Rising Sun the old, flooded workings were a menace to the working seam close below it, the Main Coal Seam, where a fall of roof could produce an immediate waterfall like that of Jesmond Dene and semi-mobile ram-pumps would deal with hundreds of tons of water a day; tubs would fill with water through their token-holes unless these were plugged and men would have to walk in water up to their knees to get in-bye. At Burradon, where for twelve months the author was a Deputy in the Main Coal Seam, the work on the coal face was nice and dry, but then some planner decided to put up a borehole to the High Main to see if it was wet; thereafter the "neuk" filler, on the longwall face, had to kneel in six inches or more of water which percolated down through the disturbed strata, finding its way to the lowest point!

CHAPTER 2 Early Years

1810

On Easter Monday, 23rd April, 1810, the sinking of the High Pit on the high ground east of Killingworth Village was commenced. The new shaft was to ease the coal-raising bottleneck at the West Moor Shaft, and also, it was hoped, to reduce the haulage and ventilation difficulties anticipated as the working distances grew over the years. The site being north of the 90 Fathom Dyke, water was expected, and a pair of levels had been driven in good time from West Moor to beneath the spot, so that a borehole sunk when the new 12 feet diameter shaft was down 420 feet could drain away the water as sinking proceeded.(1) A great deal of water was encountered at the 70 fathom, or 420 feet, level from a band of porous sandstone or "post". Oak tubbing, 9 inches thick, was applied but could not stay the flow. Eventually cast iron segments, four to the circle, were tried and these were successful.

In the event the 130 feet borehole connected with the level beneath and all the considerable amount of water duly passed down and along the prepared level according to plan, but the main pumping engine was not capable of removing it fast enough, and it began to accumulate. With some alarm the Grand Allies ordered the erection of an auxiliary Newcomen engine on the sinking pit but even this could not draw the water over and above what the main West Moor engine could handle. George Stephenson, who at that time was brakeman at the colliery, boasted to some of his, even then, many admirers that, given the chance, he could make the High Pit engine work hard enough to cope on its own. Ralph Dodds the Viewer heard rumours to this effect, sent for George and put him to work; the rest is History. The pump worked, Stephenson became Colliery Engineer, going on in due course to be locomotive pioneer,safety lamp inventor, coalowner and Killingworth's most famous inhabitant.

By 1813, George Stephenson's first locomotive was placed on the rails at Killingworth Colliery for trial, and was working by July 1814. Together with Ralph Dodds, he patented the design in February 1815.The new shaft, again split into two halves called the C and D pits, reached the coal on

A Hewer, and a Deputy Setting a Prop

The hewer has his cracket instead of knee pads; he will hew for about two hours before resting or taking a drink. The seam is thinner here and he is undercutting it. Note the round shovel used in Northumberland and Durham. The Deputy, wearing checked fustians, has a Marsault-type lamp in his teeth and is setting a prop under a "plank" with a notched axe; he is kneeling in the bottom caunch, taken to provide tub-height. Photo taken 1890's, probably South Northumberland. The Deputy would make some effort to maintain his status, hence the cap and suit, although the hewer would probably make more money. Similarly, overmen and horsekeepers might wear collars and ties; manager and undermanager would have pit-suits tailor-made.

[Bulmer and Redmayne]

12th December, 1812. The shaft was 563 feet deep.

1811

In April of this year Nicholas Wood arrived at Killingworth as an apprentice Viewer. He was sixteen years old and secured the apprenticeship through the influence of his father's landlord Sir Thomas Liddell, afterwards Lord Ravensworth. His father was a tenant farmer at Wylam and Sir Thomas had been impressed by his young intelligence. He was paid £10 per year and given £30 for board. He was a successful learner and rose to great eminence in the North East Coal Trade. He became Manager and, eventually, virtual owner of Killingworth Colliery and many other mining interests; by 1862, when he delivered his inaugural address to the Institute of Mining Engineers as President, he was the most respected mining expert in the country. Among his enterprises was the Black Boy Colliery, in Durham, of which he was the sole owner. When George Stephenson in 1818, his wealth and influence beginning to grow, put his son in Nicholas Wood's care to also learn mining, Nicholas and Robert would often call at the Black Boy public house, situated on the road linking the High Pit to West Moor Village for a meal and a drink of ale after their mornings in the mine. It is surely no coincidence that Wood's colliery was so called and that he had fond memories of his younger days at Killingworth. Robert Stephenson, incidentally, was underground with Wood in 1818 when a small explosion happened, and shortly afterwards he decided that mining was not for him; he went on to greater feats of civil engineering. He was a very good scholar and, whilst still a schoolboy, performed the mathematical calculations required to correct for latitude when his father inset a sun-dial (which still remains) into the wall above his front door.

1814

There was a fire underground and several workmen were suffocated.

Wages were good. Nicholas Wood recorded in his student notebook, now kept at the Northumberland Record Office:

The prices paid as follows: Underviewer 32/- per week, Wastemen 21/- per week, Master Shifter 3/6d per shift, Shifter 3/- per shift, Hewers 3/6d per shift (When sick *) 2/6d per shift, Master Sinker 4/- per shift, Sinkers 4/- per shift, Overmen 21/- per week, Deputies 20/- per week, Keekers 3/- per shift, Rolley waymen 3/6d per shift.

For hewing the whole coal (i.e. not extracting pillars) 4/1d. per score (q.v.) of 22-peck corves. Hewers at the screen 3/- per day of 14 hours.

* Note. They would have to be "near unto death" to receive this, otherwise they would be employed at the screens on the surface.

1815

George Stephenson was experimenting at gas "blowers" underground at Killingworth with different designs of safety-lamp. The student Wood drew the final design for him, suitable for manufacture by a Newcastle tinsmith. Within two years Stephenson was awarded £800 by the local coalowners acting jointly. The lamp was developed contemporaneously with, but quite separately from, that of Humphrey Davy. Although the "Geordie" lamp was developed from a different premise, in fact it relied on the same principle as that of Davy.

July 31st. The High Pit at Killingworth Village began to be worked by horses. Prior to this all putting had been done by hand.

November 9th. A boy was killed by a gas ignition at the very blower where Stephenson had tested his first lamp.

Tuesday, 14th November 1815 the West Moor or 'A' Pit fired. A door had been left open, short-circuiting the airflow and thus permitting a build-up of gas which was ignited by the candles of the men. Nicholas Wood, then an apprentice, noted that two bords, well-advanced and holed, were lost and three workmen were burned: Joe Grey, Backoverman; Geor. Green, a boy; and Eph. Wheeler, a rolley driver. Wheeler's father Robert had already lost three sons killed in the pits but this time Ephram, who was found lying at the bottom of the bord lived to tell the tale as did Grey, the overman. The boy Green, however, died the day after. On the following Thursday the student Wood joined a party which investigated the scene. They visited the waste and "found her in a fair state though much fallen".

Nicholas Wood was also at the A Pit in January 1817 when it again fired, this time in the West narrow bords. A wooden brattice took hold and could not be approached for the heat and fumes. It was suggested that the fire be put out by firing at it a large gun but before this extreme measure was taken someone managed to extinguish the flames with a "fire-engine" (hand-pump).

1818 - The above-mentioned gas ignition, which perhaps precipitated the abandonment of a mining career for apprentice viewer Robert Stephenson, occurred.

1819
TECHNICAL DEVELOPMENTS

In 1819 Killingworth was the world's deepest coal mine at 1,200 feet (3)by shaft and underground inclines, and by 1822 it was probably the most technically advanced. According to Nicholas Wood, writing thirty years later, a series of underground steam engines at that time powered the pit: one near the shaft bottom pumped water up a staple of 120 feet and at the same time drew coals from the same level by inclined plane over 500 yards, a gradient of 1 in 4 -(the steepest ever used later in the century would be 1 in 6); another situated 500 yards from the shaft bottom dragged coal from an increased depth of 120 feet; yet another engine at 1,000 yards from, and 240 feet below, the shaft bottom, pulled bogies from a further 500 yards inbye. The latter three engines thus pulled coal a total of 1,500 yards, at the same time raising it by 480 feet. Water was pumped by a system of sliding wooden "spears" in pipes laid down the drifts. The smoke was conveyed to the upcast shaft by brick flues.

At this time there were no cages and tubs. Shortly after his appointment as Colliery Engineer in 1812 Stephenson had introduced wheeled trams to

Backshift Hewers with a Deputy.

This photograph could not have been taken at Killingworth because the men have naked lights which were not used after 1872. However, H.F. Bulman and R.A.S. Redmayne cited a method of working a thick seam there in their "Colliery Working and Management" in 1896 and this scene would be in many respects like Killingworth. The hewers are carrying candles but the man on the right has put his down in front. The Deputy is recognisable by his cap and suit. The man behind could be a putter but looks more like a rolleywayman. The double track and stationary tubs indicate that they are on a "landing" or "flat". (flat became the name for a district.) Note the height of the seam which seems to be seven feet of clean coal. (The High Main?) The missing props to the left and the coal so far removed could mean the hewers are to set away a place there; that they are wearing their jackets means they have just arrived. Nobody is wearing knee pads. The man to the left has his water-bottle in his pocket.

[Bulman and Redmayne]

33

replace the sledges previously used to convey corves from the working places to the main roads, but hewers still filled hazel baskets which were transferred by crane from trams to rollies. In 1838 the Newcastle Journal carried an advertisement to let the "corving of Killingworth and Burradon Collieries", an early example of contracting out! Hazelwood coppice plantations elsewhere were kept a thriving industry until the sudden introduction of tubs and cages in the late 1840's, the invention of Mr. T.Y. Hall. In 1841 a man was killed falling off a rope, indicating that corf winding was still going on, and in 1853 we know from the Mines Inspector's Report that two persons were killed at Killingworth by entangled cage chains producing such a jerk as to throw them out, a third man having a narrow escape. Probably cage winding was introduced in 1842; putters were quoted as "filling tubs" then; Jarrow Colliery had cages in 1844.

Also in 1819, George Stephenson constructed a self-acting incline at the High Pit surface: full wagons descending river-wards pulled empties up by rope on a single track with a by-pass.

1822

An underground gas explosion occurred and several men were burned, included among them was Nicholas Wood.

At Burradon, then under the Killingworth management(2), the rope in the shaft broke and two men and two boys were killed. At West Moor, in the same year, several men were burnt in an explosion and some were thought by George Stephenson, speaking in 1835, to have died. It is possible that his son had been present but more likely that Robert had by then left for Darlington.

1824

During the complete twelve months the output, in chaldrons, was: A + B Pits (West Moor) 11878. C + D Pits (High Pit, Killingworth) 16444. Burradon 10283.

Assuming 53 cwt per chaldron this meant a total output for the Grand Allies at Killingworth of 102,303 tons of saleable coal.

1827

Joseph Watkin, 6, of West Moor started work as a trapper this year. He was younger than most; the average for boys starting at Killingworth Colliery was 8 years. In 1841 he was still there, putting by then.

1829

On May 13th there was an underground explosion when one boy died and several men were burnt.

1831

Nicholas Wood published "A Practical Treatise on Railroads", and became famous. A strike at Killingworth was broken by evictions, families camping with their furniture in the fields. By 1832 the Hepburn Union had collected £978-4s-6d from the men of West Moor which was all paid out to the sick, dead and unemployed. Thomas Hepburn became a hewer again and died in obscurity).

1834

A Wesleyan Chapel was opened at West Moor.

Nicholas Wood and George Johnson were appointed to report on the new tubs and winding system at South Hetton.

1835

At this time the Colliery was using candles in preference to safety-lamps in certain circumstances. There was more sense in this than at first thought appears. The candle provided better illumination, it induced more caution from the hewer and its use was confined to working the solid coal where the ventilation was better; it was barred from "broken" working. When Wood gave these reasons to a Parliamentary Commission this year he was not being as straightforward as his reputation since would imply. In fact it would have been impossible to work economically the hard coal at Killingworth then being encountered without gunpowder, and where was the sense in using lamps and then lighting a fuse to produce a fiery explosion? He implied that the candle kept the men alert to danger: by watching for a halo around the flame they would more quickly detect any gas; they would not rely on the vain security of a faulty lamp or one placed on an uneven bottom so that its gauze might become red-hot. Similarly, his reasoning was suspect in regard to the use of lamps only, in the "broken" workings (pillar extraction): here explosives were not needed, the coal being softened by strata pressure. This was just as well because the "broken" workings were always much gassier.

For at least the next sixteen years, candles were used at Killingworth in very dangerous circumstances.

1837

At Burradon a 10 feet diameter second shaft, 30 feet away from the existing shaft was sunk to the Low Main Seam. This was the beginning of a separate existence for the mine which was eventually to have a series of owners, the last being the National Coal Board.

1838

In March the Company gave two keels of coals (20 waggons) for distribution among the poor of North Shields. We can assume that this would be smalls.

1841

Wood entered into partnership with John Bowes to work Marley Hill, Co. Durham, which had been abandoned by the Grand Allies in 1815. This event marked the real start of his rise to riches. By 1849, he was living in style at Hetton House. He was still a major force at Killingworth however, and had his office on Newcastle Quayside. In 1844 he was still registered as "Viewer" of Killingworth.

About a month before J.R. Leifchild wrote his Report to the Commissioners on the Employment of Children in 1841, a man slipped off the rope at West Moor.(4) He had arrived too late for work one morning, and "obstinately laid hold of the rope, which slid through his hands and he fell to the bottom and was killed."

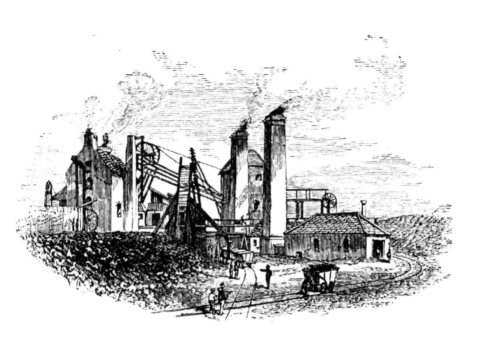

West Moor Pit 1860
Left to Right;
The Crab. A beam engine producing
rotary motion (the Crank). Just peeping over
this is the ventilation chimney. The headgear pulleys.
The screen. The winding tower with its chimney.
Chimney for steam boiler for underground use.
Locomotive shed with winder counterpoise
behind. The Crank, which gave its name to Crank Row,
drove a hemp rope which then went down the shaft for
haulage underground.

[North Tyneside Libraries]

Boys often clung to the rope, gripping it between their legs although sometimes they were let down in a corf. Usually two men went up or down together, each with a leg in a loop made by hooking the chain on the end of the rope back upon itself. They would hold on with one hand, holding in the other a stick to fend off the side, countering any oscillation. This travelling on the rope was called riding, a term which is still used for travelling in the shaft. Coming out of the pit they would be grasped by the banksman and pulled to a "settle-board" or platform and aided in disengaging themselves from the loop. A bell rang automatically when the rope was at the right height in the headframe and this hopefully prevented any overwinds. At some collieries men rode with the full corves. Leifchild reported that often many men and boys rode together, resembling a "string of onions". He did, however, mention that men were starting to travel in cages at certain collieries of the Wear, saying that it was much safer.

Leifchild also reported on the household expenditure of miners in Northumberland and Durham in this year. They were paid fortnightly for what eventually settled down to be eleven days when in full work, having the "pay-Saturday" off. However, it was more common then for a pit to work an average nine days, about two days per fortnight being lost on average while a shipment order was awaited, so the figures given perhaps err optimistically.

This loss of working days persisted for one reason or another through much of the whole life of the pits. Old miners in the 1950's still had memories of the buzzer blowing in the 1920's for the Baltic being frozen so that coal could not be exported, or timber could not be imported from there for props, or there was no empty ship to take the coal, or there was just a shortage of waggons. The fact was that the competitive selling price of coal, and the cost of producing it, were always crucial, and any slight drop in demand immediately led to retrenchment. Add to this stoppages and strikes following disputes and it is apparent that miners were never so well off as their comparative earnings implied. They earned more than agricultural labourers, of course, often to the annoyance of landowners, but less than ironworkers at Walker, in the same parish as Killingworth, although rather more than a plate-mill furnace man there, with a wife and two children, who earned 18s per week. Leifchild conjured up a typical mining family, from Durham, of father and three working sons, aged 17, 12 and 8 earning between them £5-3s-7d per fortnight, whose total outlay, including beer money, benefit funds and conveyance of free coals came to £4-9s-0d. Notable among their purchases were 28lbs of meat (mutton and bacon) and a stone of soap. The wife would bake all their bread from a stone of flour and a further 3 stone of "maslin", a mixed grain, some of which may have been used for broth. House, as well as coal, was free, but 6d. per fortnight was spent on candles for the house (16 per lb). This calculation was similar to that of Mr. Micawber and the result, supposedly, happiness, but it depended not only upon an iron will of the housekeeper but also upon the colliery actually working the full fortnight. Five years later the men at Killingworth were averaging nine days but often working only eight between each pay.

There was a compulsory holiday of two weeks at Christmas at this time, unpaid. The other holiday of the year was Easter when Good Friday was taken; Easter Monday had also occasionally been declared a holiday.

FOOTNOTES

(1) This in itself indicates no mean surveying achievement for the time. The Vernier Scale and Logarithms, invented long before Killingworth was sunk, had led to the development of the Theodolite. Scientific-instrument makers tried to produce ever more accurate devices to measure horizontal angles. By averaging a series of readings of the same angle obtained by swinging the telescope in both directions and then repeating these measurements after vertically rotating, it any small errors could be compensated for, any readings outside a predictable mathematical range being discarded. An angle could thus be measured accurately to a few seconds of arc with an early Theodolite and even without a telescope angles could be measures over open sights to a minute. Multi-place logarithms were then used to ally these angles to distances to produce co-ordinates. Without two shafts, of course, no check was possible and the starting base-line was only two wires in the shaft. Had there been an underground connection with another pit the surveying would have been easier, but there was not. A magnetic compass dial would have been reasonably accurate for mapping underground workings, iron rails being absent during the first years of the mine, but hardly sufficient for so precise a connection. The correlation would depend on two weighted strings hanging down the shaft and about a mile of theodolite traverse underground (and the same on the surface as no Ordnance Survey Map existed then!)

(2) At this depth the rock temperature would have been 70°F. The air current, furnace driven, being much less than that considered acceptable nowadays when electric fans are used, would have a cooling effect but this would be more than offset by the presence of underground steam boilers and smoke flues and the heating effects of men, horses and the oxidation of coal. The working temperature, which would persist throughout the year, was then, warm but acceptable. But shortage of oxygen, and air-currents insufficient to flush away firedamp and humidity, were other concerns.

(3) Burradon was worked from 1820 by the same Company. The NCB-published date of its commencement, 1837, is obviously taken from the Record of Borings and Sinkings recording the Engine Shaft. There were two shafts, each of 10 feet in diameter and one of them pre-dated the one

recorded by many years; no doubt it was a partitioned shaft like Killingworth.

(4) A contemporary account kept by Mathias Dunn, then a mining engineer, graphically describes a similar accident at Wallsend. A man slipped from the rope, hitting two boys lower on the rope as he fell and almost dislodging them. He fell all the way to the bottom where his head and body disintegrated so badly that a coffin had to be sent down to gather up the brains and intestines.

CHAPTER 3
"Having Amicably Agreed With Their Masters"

By 1841 a Mr. Collings was Viewer of the Colliery, living at Killingworth Hall which still exists. Killingworth House, higher up the hill and on the opposite site of the road, the former property of a Georgian admiral was a much grander, double bow-fronted edifice which later became the home of the last General Manager of the Colliery. It was demolished in 1956.

1842

The total number of men and boys at Killingworth and Burradon together was 411, of whom 111 were under 18, 43 of these latter boys being under 13. The wages they received were:

Overmen 25s - 32s per week. Back Overmen 22s - 25s per week. Deputies 3/4d per day. Hewers 3/9d. Putters 1/3d per "score", the average wage at Killingworth being 2/6d per day. The average wage at Hetton, a sister colliery, 3/9d per day.

When a worker was laid off because of an accident, "short of death" he received "smart money" at the following rates, per day: Deputies 2s0d. Hewers 10d. Trams 8d. Rolleydriver 4d.

Leifchild in his Report noted that the miners quickly recovered from burns and scalds, being small men of spare build. The above term "tram" needs here some explanation. A fully-grown putter was known as a tram. Younger boys often worked in pairs, the elder being a "headsman" and the younger a "foal" and would be paid out of each 1s. jointly earned 8d. and 4d. respectively. Two boys of equal age were called "half-marrows". "Helpers-up" were stationed on gradients and literally did donkeys' work. (13 real donkeys were kept for the very stiff gradients; ponies were not introduced until later in the decade.) All the putters worked 12-hour days.
By 1842 there was a school at West Moor run by the Colliery charging 3d. per week for a whole family.

At Hetton, also under Wood's control, small ponies were being introduced. A minimum age of 8 years for trappers was instituted.

1843

On 25th November the "Newcastle Journal" reported the fining of five shillings each of eight boys from West Moor for throwing stones, down the pit, at Henry Laverick a Deputy. This does not appear to have been a childish prank as at least one of them was seventeen years old and was probably the result of some grudge. Laverick was sitting behind a brattice when he was pelted. In the Court John Gilroy, Overman, identified the lads, despite the poor light, knowing that they were away from their posts, as J. Cornwell, W. Thompson, J. Saint, M. Story, A. Carr, J. Hand, J. Dodds and M. Poat. Five shillings at the time represented two days hard labour putting, then by hand, the colliery having only thirteen donkeys for the steeper gradients, ponies not yet being used. Possibly they were working half-marrows "helping up" in which case their fine equalled four days' wages. Another member of the Hand family was to be in trouble during the strike of 1849. Fore-overman Gilroy was to give evidence in Court a year and a half after the stone throwing incident on a much more serious matter.

1844

The Report to the Commissioners of this year gives employment figures of 303 for Killingworth on its own, consisting of:

105 Hewers. 85 Boys under 20 years (underground). 30 Overmen, Deputies, Wastemen. 27 Surface (Banksmen, Brakemen, Engineers etc.). 24 Carpenters, Smiths, Masons etc. 12 Boys under 20 years (Surface). 20 Shipping, etc.

Tonnage laid out (see glossary) averaged 0.550 tons per man per day. This figure seems unbelievably high and there must have been seething discontent among the Hewers, but the figures were given to Tremenhere, the Reporter, by the management, as the average from 5th April 1843 to 5th April 1845. During this same period there were strikes and a self-imposed earning restriction by the men and after this began, in May 1843, Hewers' earnings dropped to 2s 10d per day.

On January 18th, 1844 there was an underground explosion and five men were killed. The accident may have resulted from an overheated safety lamp not hanging but standing on its bottom, sheltered from any draught, closer to the work of a Hewer. There were sixty men and boys in the mine but the explosion was confined to one district and the majority escaped unharmed. Work continued elsewhere in the mine until word reached the other districts and even then there was no rush to the shaft. About this time it became the general custom for all the miners at a pit to leave whenever there was a fatal accident, and that has been the practice until the present day.

The men killed were:

John Storey, William Hardy, John Nicholson, William Richardson, Thomas Bates.

The first three each left widows and children. A young man Joseph Hindmarsh, was badly burnt but seems to have survived.

The low wages resulted in strikes across the coalfield. The Marquis of Londonderry, after twelve weeks of strike at Seaham, began employing men from outside Northumberland and Durham. He wrote an open letter saying he had brought in 40 men from his estates in Ireland and would bring in 100 more if his workmen did not return to work.
At West Moor another long strike was broken by evictions. Nicholas Wood made arrangements to import 400 Scotsmen if necessary.

1845
On Thursday, April 3rd a serious explosion occurred with this time 9 men killed.

It was afterwards claimed by Nicholas Wood that the explosion resulted

44

from an ignition of gas by a candle carried by John Hindmarsh, a trapper boy, who ran ahead of the safety-lamp-carrying hewers as they walked in bye. Wood said a blower must have burst out beforehand and that the lad ran into the cloud of gas.

From our present point of view we can see that the fact that there were any candles at all was foolhardy. We also know that the acceptable volume of air per minute circulated in a mine at that time was about one tenth of that deemed necessary nowadays. But the argument for candles persisted. Coal-owners produced varying reasons for their use: better illumination; better vigilance; the men would remove the tops of their lamps anyway for a better light despite threats of three-months jail sentences; they would smoke anyway; the mines could not be worked without gunpowder and accidents were inevitable.

Two hewers had just been drawn out of the pit, and their shouted warning was the first intimation of the disaster to the men on the surface. They had felt the explosion just as they left the shaft bottom. A large silent crowd of men, women and children soon gathered around the pit head and, as was usual in such circumstances, volunteers came forward to attempt a rescue. A stream of water, also usual in such disasters, was poured down the shaft to give air to the men descending. The rescuers made their way into the mine, some collapsing and needing resuscitation as they went, until at five to six hundred yards they found two dead boys and two injured men whom they got out alive. Further in they found another four bodies. Four men were still missing but their bodies were found a few days later after the inquest was opened.

Those who died were:

John Sharp, hewer, left a widow and six children.
William Sharp, his brother, also a hewer, left a widow and seven children.
Robert Hall, deputy, a single man on the point of marriage.
Mathew Thompson, putter, a youth.
William Moulter and Thomas Stewart, trapper boys.

Thomas Thompson, a hewer, left a widow and four children.
Peter Tweddle, hewer, a young man.
John Hindmarsh, a trapper, and John Grey, a putter, both boys.

The last four were the bodies found later. Probably the Thompsons were related, if not the same family.

The inquest was opened by Stephen Reed, the Coroner, at the Colliery Office on Friday, the day after the explosion, the six bodies being laid out for the inspection of the jury while the coffins were made at the joiner's shop. The jury were not pitmen. Like all non-mining people of the day Reed regarded the miners as people of low intelligence. Stephen Reed was of the stuff which makes villains in Victorian melodrama. The previous year he had refused the miners' representative leave to question witnesses at the Coxlodge Explosion inquest. He was a successful lawyer on the make and his early main trade had lain with the coalowners, for whom he had sycophantic reverence. Earlier in his career he had acted as a patent lawyer and by 1838 owned the rights to the use of an improved hook with a moveable tongue which could prevent accidents from falling corves in shafts. In letters to the "Newcastle Journal" he sought publicity and at the same time warned against parties using his patents without licence. In 1850 Reed's Patent Iron Block Sleeper was advertised. Later he would be criticised publicly by Mathias Dunn, the first Mines Inspector, after the Walker Colliery explosion in 1862, where coal had been worked by gunpowder while the ventilation furnace had been reduced by one-half for works in the shaft. Reed had opened with an uncalled-for speech praising the "ventilation and scientific management of Walker Colliery" and had also solicited the irrelevant opinion of the furnaceman. When the Institute of Mining and Mechanical Engineers was formed he was an unlikely Committee Member, but he was not an engineer and would have little in common with the others; he soon disappeared from the rolls.

Reed opened the inquest early, he said, for the sake of the widows and families who would have had to live with the bodies until the following Tuesday. He was just able to fit the hearing into a slot before the inquest on a woman who had poisoned herself in North Shields. It was 4.30 and clearly he was hoping to make a quick job of it. He substituted his liveried coachman for a missing juror; he told the half dozen reporters (1) present that they were there under sufferance and that they must not take notes of his every comment, and ordered them to erase what they had written. He stopped Mr. Wales, the underviewer, from describing the technical details of the ventilation: they were unnecessary as the ventilation was known to be excellent. There was no blame due to the owners. No doors had been left open. He would have wrapped up the proceedings there and then, but, because of the yet to be recovered bodies and a juryman speaking out, the inquest was postponed until the next day. Hugh Taylor, a respected Viewer from Earsdon, was called to testify but the conclusion was reached that it had been an unpreventable accident.

What happened to the widows and orphans is not known. They may have lived on in their Colliery houses for a while but pressure for housing for new men would make their position increasingly untenable. Hearsay evidence is that widows worked in the fields for farmers until their eldest sons could provide for all, otherwise it was the workhouse. Collections were taken on the following Sunday in the local churches when £26-0s-7d. was donated, £11-15s-7d. coming from the West Moor Schoolroom which then doubled as a chapel at weekends. We can only imagine the trauma which followed these disasters, the stunted lives resulting from broken childhood, the grief of the relatives and the anguish of working mothers with small children.

By the time of the Burradon Explosion in 1860, and over which he presided, Stephen Reed was boasting of fifty years as a Coroner, a claim not strictly true, he having still had a private practice as recently as 1838, when he no doubt acted as Coroner when required. He lived in a splendid house in Queen's Square, Newcastle. By 1844, he was described as the Northumberland Coroner and seems to have had a busy time of it. He was not alone: shortly after the Killingworth explosion of that year there was a much worse disaster at Jarrow, where the Durham Coroner J.M. Favell again prevented the miners' representative, Martin Jude, from questioning witnesses, and on receiving a verdict of Accidental Death asked the jury, "By that you mean that the owners and managers were not to blame, and that the workings of the mine were in proper order?" "Just so," was the reply. Ellen Wilkinson(2) cynically said that this meant the explosion was due to the inexperience of the workmen!

Also in 1845, a Professor Graham demonstrated to The Chemical Society the properties of mine gases. One of the samples he used he collected, himself, at Killingworth. It came from a jet, issuing from a stratum of sandstone, which had been kept uninterruptedly burning as the means of illuminating a horse-road for upwards of ten years!

1846

Writing on the state of the mining population, Government Reporter Seymour Tremenheere said of Killingworth and Burradon that there was a population of 2,500 belonging to the works. He wrote, after visiting Nicholas Wood:

"It would be difficult to meet with a labouring population of a similar kind, better off as regards wages, habitations, gardens, and the ordinary requisites of physical well-being proportioned to their class. The management is conducted with great liberality. Gardens are attached to all the houses, and each has been fenced in and separated from the rest, chiefly at the cost of the owners, who supply wood or old staves for the purpose. This is a great promoter of the neatness and careful cultivation that prevails in them. Although the Company have a farm of 1000 acres in hand, and keep 100 horses, they nevertheless give to the colliers as much manure for their gardens as they require. The agent informed me that he estimated the value of the manure thus applied at £200 a-year. The manure of which the farm is thus deprived is supplied, by purchasing from the town. They have house and coals free. Seven waggons are sent to Newcastle on market-days to bring home their purchases.(3) The owners subscribe a handsome sum yearly to the colliers' fund, for sickness; and in cases of accident, they also pay the doctor. The comfort and health of the population has been studied by the formation of a deep brick drain, covered at the cost of £300, through which the engine water is pumped to keep it clear. A steady man will earn from £50 to £60 per year, exclusive of the earnings of his sons. There has been a school at the colliery 40 years, to which the owners have contributed."

However, John Wales, the under-viewer told him that when the men were on short time, working only eight days a fortnight, they became fractious and ready for strikes, especially the younger ones:

"Our men remained on strike ten weeks. We got the 'deputies,' waste-men, and waggon-men to hew; the rest did not molest them, except by shouting at them. We were obliged at length to turn many out of their houses; we got six policemen, but two would have done, as they made no resistance. They remained out of their houses six days; they got deeply into debt during the strike, some £10 or £12, and are not clear yet. Many had been saving money in anticipation of the strike for ten or twelve months beforehand.

"I have known this village all my life; the people are much improved and bring up their children better; most of them attend some place of worship. Cock-fighting, gambling, fighting, &c., are nearly gone by. We have had four of the county police in the parish and they have had a good effect in promoting better habits, in putting down card-playing and gambling in the public-houses, & c.

"Most of the young men give their earnings to their parents till they are about eighteen or twenty, the parents allowing them 2s. or 2s6d. a-week pocket-money. They do not hew till they are 22; they save a little money and marry at about that age."

Tremenheere also visited the school, which had mixed and girls-only rooms with 84 and 40 pupils respectively, and an evening school attended by 27 young men. There was a library attached. For the last two years the Vicar of Longbenton or his curate had held a weekly service in a schoolroom. At the time though, the miners generally preferred the dissenting chapels

Michael Anderson, Coal Hewer, and family. Michael was just about the last lad to be taken on at West Moor. He was soon transferred to Dinnington, travelling each day in "the carriages", (ex-main line) where he was putting to his uncle John Todd when the latter was killed by a fall of stone in 1886.

James Thompson, Coal Hewer of Killingworth Village, with his only surviving son, also James. His two earlier children both died, one of diphtheria, the other drowned in the High Pit pond. His wife is Mary, ex-Todd, nee Anderson. Mary lost her first husband and one of their sons to underground accidents.

[Thompson Archive]

of which there were by now three: Wesleyan, New Connection and Primitive Methodists. The latter, also called "The Ranters", had political overtones and attracted the young firebrands. Conversely, the Wesleyan was conservative in outlook, attracted steady old hands, deputies and the like, and kept a library. The New Connection which must have been, in outlook, somewhere in between, was noted for its music and singing. Fifteen years after Tremenheere's visit, 24 year-old Jane Anderson, living in Crank Row, married to a Catholic Irishman who forbade her to attend Protestant chapels, would surreptitiously leave the house door ajar on Sundays to listen to the hymns from the New Connection Chapel wafting across the pantiled roofs of adjacent Cross Row.

1847

A Reading Room was built for the workmen by the Company and on Christmas Day a "numerous party of workmen" dined together there and "spent the day in social enjoyment and improvement." (Later Management was to regret the sparse use made of the Reading Room by the miners as against their other preferences, viz. Beer and gardening.) The Reading Room, an imposing single-storey stone building, stood until 1946 and served as the local Mechanics' Institute.

The lessees of the Colliery were now referred to as John Bowes and Partners, and this was the name of the Company until it closed. Nicholas Wood became part-owner.

1849

From the Newcastle Journal of 7th July, 1849:

"The pitmen of West Moor Colliery, who have been on strike in consequence of a dispute about wages, having amicably arranged with their masters, have resumed work. If they had not done so, they would have been turned out of their houses last week."

This short mention of the Colliery, which appeared on its own and without further editorial comment or explanation, is heavy with irony when seen through modern eyes, but at the time was probably noted with complacent disinterest by most of the readers. It masked a saga of bitterness and struggle which had gone on for years between men and masters, each side becoming increasingly determined, as they felt the polarising influences of the rebellions which had swept Europe in 1848. Within three days of the notice appearing, serious riots occurred at Killingworth.

The general strike among the pitmen of the Tyne and the Wear of 1844 had failed, the men being literally starved into submission with their Union funds exhausted and their credit suspended. In 1849, however, they had adopted the plan of striking at one colliery at a time, the men from the other working pits contributing money to those who had stopped work. There had been strikes at Seaton Delaval and Hartley and it had been lately the turn of Killingworth. After some weeks of stoppage the owners had arranged for a small detachment of police to be stationed at West Moor to protect any men who were willing to work. It is not known for certain if "black-legs" were brought in, but this had been the case elsewhere, and in an earlier strike at Killingworth the officials and craftsmen had been put to hewing, successfully, keeping the colliery viable. But, in any case, the men saw that they were being injured more than the proprietors, and the threat of eviction seems to have been the final push which drove them to violence.

Miles Hand was arrested for assaulting Robert Cooperthwaite, a strike breaker, and was given three months' hard labour by a Judge who admonished him "for interfering with the great liberty of a person to sell his labour freely." ("Newcastle Chronicle", July 20th.) The real riots, however, happened next and the culprits had to await the Assizes for their trials.

On the night of 9th July, over two hundred pitmen gathered at West Moor, wearing disguises and masks, and marched to the shaft where they cut the cage ropes and threw tubs and whatever else they could find down the shaft, after the cages. This effected great destruction and put a stop to the works. The small body of police and colliery watchmen were driven from

Killingworth Station 1910
[Norman Whitelock]

their posts and collected at the colliery office, where the rioters showered them with stones, breaking the windows and destroying the building. The occupants then retreated to an adjoining cottage, where they were again assailed. Sub-Inspector Jenkins, who said afterwards he had heard two guns fired, escaped by a back window and hurried to Newcastle to inform Superintendent Stephens, who quickly mustered a large force of police. As many men as possible were then crammed into a large omnibus drawn by four horses which was soon galloping along the road to West Moor.

Meanwhile, it appears, the rioters had visited Burradon, where they had also done considerable destruction. They must have returned just in time to meet the police, who were armed with cutlasses. The men's courage failed at the sight of the disciplined force and on Stephens' command of "Draw swords!" they broke and fled across the fields to Killingworth Village. Stephens then ordered a charge. The police constables were described with admiration in the "Newcastle Journal":

"It being of course, desirable to secure the leaders, orders were given to "chase" and gallantly was the order responded to, the most active of the policemen clearing the hedges at a bound, the others pressing forward, and all displaying a degree of tact, discipline, and alacrity truly admirable. In vain the fugitives separated so as to divide the attention of their pursuers, for the latter spread out and covered the ground making a long sweep across the country so as to render escape almost impossible. Police-constables Joseph Pegg and John Marshall, being the fleetest in the force, outstripped their brother officers, and each of them secured a prisoner a little beyond Killingworth: Thomas Little pounced upon another prisoner skulking along the lane leading towards the village, and other prisoners were taken at various points."

Somewhat less gallant, was the simultaneous bursting into houses of the colliery rows, to haul out suspects, whose wives were tearfully protesting their innocence. In all, six prisoners were escorted back to Newcastle, reaching the Prudhoe Street Station at half past six in the morning whence they were removed to the Moot Hall. They were William Arnott, 29, hewer, the alleged leader, Mathew Teasdale, William Puncheon, David White, John Buxter and Henry Walker.

They were taken that afternoon before Magistrates, Captain Potts and Aubrone Surtees, Esq., when they were remanded in custody. Eventually, all except White, who was replaced by the newly arrested William Coulthard, were sent to the next Assizes. Puncheon was also released soon afterwards. Superintendent Stephens told the Magistrates that he had received a letter from Mr. Johnson, the Viewer of the colliery, requesting the aid of the military, saying that he was apprehensive of another outbreak. Evidence was also given that riots had occurred on the 18th, 23rd and 25th June, the parties usually assembling at midnight when they marched through the village, using threatening language and menaces, in order to deter the men from going to work.

The Assizes took place, mercifully soon afterwards, and on 2nd August the Newcastle Chronicle reported that the five men had pleaded guilty on the advice of their Counsel, Sergeant Wilkins. The Hon. Mr. Liddell(4) on behalf of the Owners said that his clients had no ill feeling towards the men and that they would withdraw the prosecution. The Judge said the prosecutors had been lenient, and discharged the prisoners on recognizances of £40 each, to appear to receive judgement if necessary, with the understanding, that if they behaved themselves well, they would hear nothing more about it.

What the true story behind this move was can only be conjecture; the whole legality of it seems strange from a modern perspective. The misery of the men during their three weeks in jail, with no money, anticipating transportation, at the least a prison sentence and certain eviction for their families, can be imagined. Did the owners consider they had suffered enough? More likely, it was just their desire to lower the tension and keep the pit in production to cover their fixed costs as lessees. If this really was the explanation, the episode was yet another example of riot winning for the workers what their as yet non-existent votes could not. It also showed the extent to which the legal system was in the hands of the controlling class.

Two years later, ringleader William Arnott was still employed at Killingworth, according to the 1851 Census, so things must have settled down. By that time the Mines Inspectorate had been established, but a much deadlier incident that the riots was to occur: the last major explosion at the mine.

This year Nicholas Wood sold off Burradon, which was then no longer under the control of John Bowes and Partners.

1851

October 31st saw the last large explosion at Killingworth. Nine men were killed. Six feet, of the eight feet thick High Main Seam, at a very considerable angle, was being worked 190 fathoms from the surface. At that time the whole colliery was ventilated by 30,000 cubic feet per minute of air, good for the times, driven up the upcast shaft, by two furnaces, one at each level worked. The air was split four times and the district in question received 6,000 cubic feet per minute. Leading headways 20 yards apart were being driven, holed at intervals of 40 yards, which occasioned 60 yards of bratticed air for each heading. There were 130 men in the pit at the time, 14 in the fatal district. One of the men had taken in both candles and gunpowder. During the investigation it came out that the ventilation had been much deteriorated by the extreme wetness of the upcast shaft, which neutralised the effect of the furnaces; since the explosion a proper repair had been made, and the ventilation improved.

The previous week and a half had seen two separate ignitions of gas with men burnt. "Lows" or gas-caps had been observed by hewers, on their lights; they had been told to keep their candles down, away from the roof. It was claimed that the men were given another sixpence per score if they worked with safety lamps, but that they were not necessary in the "whole" working. A man working with a safety-lamp said that he was frightened to be in a gassy place with lads on the haulage plane sixty yards away using candles. The story was confused, but it seems that the explosion happened during a haulage stoppage, and hewers were idling, waiting for tubs. They may have been smoking, but a witness could not be induced to admit this.

Really, it was the old story: naked lights allowed in some parts of the mine. (Naked lights were allowed in parts of mines well into the 20th Century.)

Those who died were:

George Grey, 56, leaving a widow and a boy.
Robert Carr, 51, leaving a widow and 4 children.
George Campbell, 28, leaving a widow and 3 children.
George Hewitt, 51, leaving a widow.
Jacob Hedley, 36, leaving a widow and 3 children
 Owen Hay, 36, leaving a widow and 2 children
 Robert Jordan, 11.
William Mason, 13.

Of the other four or five who were badly burnt, one died later.

FOOTNOTES

(1) One of the reporters commented on not understanding the very strong accent of the village.

(2) "The Town That Was Murdered", Ellen Wilkinson MP, 1939.

(3) This seemingly innocent comment may have disguised an iniquitous "Tommy Shop".

(4) Probably Mathew Liddell, a consulting mining agent of Benton Grange. He was possibly a relative of Lord Ravensworth. As well as acting for John Bowes and partners in this case, he also served the Earl of Carlisle at Killingworth; in addition he acted for the Borough of Newcastle at £62 per year. (He was also Chief Viewer for Coxlodge Colliery.)

CHAPTER FOUR A Victorian Mine

1851

Mathias Dunn criticised the method of working whereby "all props are set by overmen and deputies and it is not left to the judgement of the collier how or in what manner to apply them." This was the first airing of an opinion, which later grew to a clamour, and eventually produced systematic timbering (q.v.).

1852

On December 13th a man was killed in the shaft.
By this date the High Pit was entirely upcast, but some coal was still being produced.

1853

On April 11th two men were killed by the chains which connected the cage to the rope becoming entangled, so as to produce such a jerk as to throw the two out whilst a third "had a very narrow escape." It is likely that some form of spring-loaded catch or "keps" (q.v.) was in use by that time and the cage resting on the keps allowed the rope to slacken and the chains to snag. This could have resulted in the cage tilting or being severely jerked.

On April 16th a man was killed by a fall of stone.

1854

During this year 3 men were killed:

1 in the cage in the shaft
1 by a fall of stone
1 by rollies.

Technical Developments

In 1852 experiments were conducted by Nicholas Wood on the use of steam jets to produce an upward air-current in a shaft, and thus ventilate the mine. He chose three of the collieries over which he had control: Hetton, Tyne Main and Killingworth.

At that time Killingworth had two furnaces at different levels. Wood used the partition of the main West Moor shaft not normally upcast, and a suitable arrangement of underground doors, to conduct his experiments. He was conscious that at that time 250 lives were being lost annually in British mines from explosions alone, and spared no expense. Holes were drilled through the wooden partition brattice at intervals, from top to bottom, for pressure-difference readings and the steam jets installed at the bottom. The jets were powered by three large surface boilers, strenuously stoked. Exhaustive readings were taken, but in the end Wood concluded that the system was inferior to the furnace; perhaps he felt this from the start because he had had a short experience of a steam jet in Wales in 1828, which had been unsuccessful, but at least he proved that there was no future in the system, and development of it was curtailed throughout the coalfield.

In 1853 Wood continued his scientific work at Killingworth, developing a Stephenson lamp using gauze instead of perforated plate, which was tested at a blower. He commended this lamp for its property of extinguishing itself in the presence of inflammable gas.

By 1855 it would seem that the Newcomen type pumping engine had been abandoned in the main shaft. The water was wound in tubs in the cage for 12 hours per day. The underground haulage engine was powered by three boilers which, by their dimensions (5 ft. 1 in. diameter, length 34 ft.) seem to have been those used for the steam-jet experiment. The boilers were situated 150 ft. from the shaft and a 10 in. pipe led steam 690 ft. down the shaft and then 36 ft. inbye to a receiver, then back to the shaft again through a 5 in. pipe and then down a further 360 ft. and 34 ft. inbye to a second receiver and thence 35 ft. to the engine. The entire distance was 1,341 ft., the engine being 1050 ft. below the level of the boilers. Any heat loss made the system inefficient so the pipes were wrapped in an inch of felt covered in asphalt. The receivers or reservoirs of steam were 13 ft.

Locomotive from Stephenson's time. Still in use at West Moor in 1872.
[North Tyneside Libraries]

x 5 ft. 2 in. diameter and 5 ft. x 3 ft. 6 in. diameter. The exhaust steam from the engine was discharged to the upcast shaft at the 690 ft. level. The main pipes to the first receiver went down the upcast shaft but the next, smaller, pipes went down the downcast up which the pit water was being drawn and so were cooled a little, the engine thus losing some power.

This engine drove a main and tail rope haulage, 700 yards long, on a near-level gradient, formerly a horse road. Another engine worked from the surface beam engine formerly used for drawing coal, it drove a hempen rope which was taken 1050 ft. to the shaft bottom and then horizontally for 800 yards, there to haul tubs up an incline by direct haulage. The empty set free-wheeled down by gravity, taking the haulage rope with it. This rope (unlike others below ground) must never have been replaced by steel: a newspaper reporter commented on it in 1878. Its 8 ft. diameter drum and crank gave Crank Row its name.

1855

General Rules were drawn up and posted at the pit-head by the Manager, Mr. Johnson. Included among the minimum standards required by the Inspectorate were special rules for the Colliery. Included were:

The person who keeps his safety-lamp in the best order for a period of 3 months will be entitled to a premium of 5s. and the second best to 2s.6d.

No swearing or fighting is allowed in the mine.

...... if, in the absence of a deputy, any place requires additional timber, the hewer is not to continue his work, but to leave his place and to find the deputy for the purpose of having the requisite timber set.

These rules continued in use until the Coal Mines Regulations Act, 1872 was passed in Parliament.

By 1869 a prize was still available for the best kept lamp but every work-man had to take the gauze home every day to clean it himself and take it to the workshops for any repair. In practice men took the whole lamp home with them.

1857

This year the lessees at Killingworth Colliery began negotiations for the purchase of Gosforth Colliery, which was adjacent. When these were completed successfully, the two Collieries were interconnected underground, and Stephen Campbell Crone moved from Anfield Plain to become Manager of the combined mine. Killingworth High Pit, being no longer used to draw coals in quantity, had become the only upcast shaft and the West Moor Pit had lost its furnaces and was the sole downcast.

Also this year Joseph Straker of Tynemouth bought Burradon Colliery for £29,800.

1861

The Census for this year shows Mr. Crone living in some style with his large family at Killingworth House, where he stayed almost until his death in 1898, although by 1881, after his wife had died, the huge house had been split into three parts, two of which were let to local bourgeoisie. Crone became highly respected in professional mining circles, and was a founder member of the Institute of Mining and Mechanical Engineers, at Neville Hall, Newcastle.

Meanwhile the catalogue of fatalities continued:

May 27th. 1 killed by coal waggon.
June 13th. 1 killed by fallen stone.
July 22nd. 1 killed on engine plane by a water tub.
November 27th. 1 killed on Killingworth waggonway.

1862

February 12th. 1 killed by crush of tubs.
August 6th. 1 killed by coal waggons.
August 9th. 1 killed by fall of stone.

August 12th. 1 killed on an incline underground.
October 29th. 1 killed on engine bank.
December 18th. 1 killed on the waggonway.

This particularly bad year must have resulted in a general tightening up:

1863

June 27th. A boy going home from work, playing on the tops of loaded waggons, slipped and died later in Newcastle Infirmary.

1864

May 5th. 1 killed on engine plane.

1865

March 30th. 1 killed by crush of tubs.
April 21st. 1 killed by fall of stone.
September 12th. 1 run over by engine.
November 9th. 1 run over on waggonway.

1866

May 27th. 1 killed by bursting boiler.
June 25th. 1 killed by fall of stone.
October 15th 1 killed, run over by tubs.
A bore-hole from near the shaft-bottom found the Harvey Seam at 1265 ft. below the surface with "good cinder coal".

1867

This year there were no deaths. Only Mathias Dunn, H.M. Inspector of Mines died.

1868

Again no reported deaths. Safety legislation was paying off.

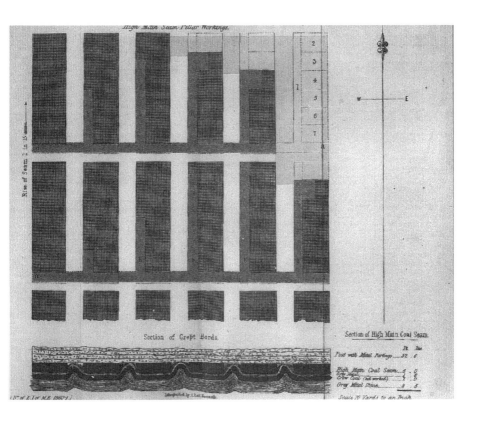

Plan of 1860's pillar extraction of very valuable High Main coal left from early days.

[Transactions N.E. Institute]

1869

A fall of stone killed Thomas White, 84, and Peter Carr, 13. The old man White was under the charge of Robert Langlands, Master Wasteman, and was repairing a roadway used only to convey coals to the ventilation furnace. Before setting the wastemen to work, Langlands "jowelled" the roof, hitting it slowly with the flat end of a pick and, finding it "rather heavy", told them to take care. Hardly had he left when some stone fell. Word was sent to recall him, by means of a pony driver engaged in transporting the selected quality coal for the furnace, and he came back to examine the roof once more. He pronounced no more would fall - he afterwards said that he "would not have been afraid to work under it himself" - and left again. Pieces of rock again began to fall, and the "poor old man White" was advised by the two boys, with only a tiny fraction of his own experience underground, to come away but he persisted, and a very large roof fell, killing him and the boy.

In his Report the Inspector strongly criticized Langlands but there was no direct evidence against him at the Inquest.

During the 1860's the people of Killingworth (in 1861 the population was 1,786) knew only too well of the terrible disasters happening elsewhere. In 1862 there was an explosion at Walker, (in the same Parish of Longbenton) and the terrible disaster at the New Hartley Pit, at Seaton Delaval, where the pumping beam and its attached spears fell down the only exit, carrying away the brattice, and entombing all the miners below ground at the time. After the Hartley disaster the Union collected information on shafts and commented on West Moor having four cages, beams and rods, and pipes all in one 14 ft. shaft.

But it was the 1860 Burradon explosion which effected them most; many would have relatives, and some would work there, even though the pit no longer belonged to John Bowes and Partners. Burradon had for several years been connected to, and worked together with, Seghill Pit, but recently the colliery had been sold as a separate concern to Joshua Bower of Hunslet, near Leeds, to become part of the Coxlodge and Burradon Coal Co. On

The Disaster at Burradon, 1860

From a photograph taken shortly after the explosion. The engraver has confused the detail of the headframe, probably because of the faint quality of the original; this being so then the objects in the foreground could be coffins and the Pickwickian gentleman possibly Stephen Reed directing operations. Two more top-hat wearers can be seen among the lookers-on. Are they the Coroner's men making enquiries? More watchers are waiting up on the heapstead. The construction pointing skyward, centre, is the screen. The square brick tower to the left of the pulley wheels is built directly over the upcast shaft and the fumes from the underground furnace can be seen. Extreme Right of the picture are full chaldrons on their way to West Moor.

[The Illustrated London News Picture Library]

March 2nd at 2.30 p.m. an explosion at a hewing place sent the agile haulage boys running for the shaft bottom where Thomas Alderson, the back overman tried to send them back inbye, saying that the danger was now over. They ignored him and were saved, Alderson himself died. William Kirkley, the fore overman, was told in the office and immediately descended the mine, met fifteen lads at the shaft bottom, and took some of them back inbye, including William Urwin, aged 14, by the scruff of the collar, when a second explosion was heard, and they all ran back and were saved. Urwin had been talking to Ben Nicholson, deputy, father of seven girls, when he heard the bang. Ben died at the place of the ignition. The men were horribly mangled, and had to be brought out by the rescuers in blankets, and even coffins were taken down to collect the severed torsos in the end. There were many more men, however, who were at first unharmed, even after the second explosion, but when they left their district they passed through the fatal fumes left over, and were suffocated, or poisoned. In all 76 men died.

The Inquest was a lengthy affair, presided over by Stephen Reed, and revealed a scandalous state of affairs. Men had complained for months about the poor air and ignitions of blowers. The "whole-coal" was worked by naked light, with safety-lamps only in the "brokens". Hewer William Dryden even walked out, and took a job at West Moor, prophesying doom any day. Part of the pit had been laid off only ten weeks before for gas accumulating in a goaf. There was only one barometer between Burradon and Coxlodge to tell when low pressure might cause the gas to expand. Men had been travelling in the dark over a particularly dangerous stretch of roadway rather than risk igniting the gas known to be there. The connection with Seghill had adversely affected the ventilation. For a while, the Underviewer had shared his duties between Burradon and Coxlodge, a long way by horse. Mr. Carr, the Manager, had a portion of his salary dependant on output. Worst of all, however, William Williams claimed that two blowers had ignited at Maddox's bord, the seat of the explosion, that very morning!

The Coroner allowed the issue to be clouded by technical discussion on a fall of stone blocking the ventilation and whether a door might have been left open too long, and allowed fierce cross-examination of witnesses, among them Mathias Dunn, the Inspector, by the Owners' representatives and summed up without any criticism of a deplorable state of affairs. He made his usual short speech, and referred to the degree of mystery which attended these disasters, offering the jury no advice. They returned a verdict of accidental ignition of gas, and said that it was probably caused by the fall and the door left open. Reed refused to accept this suggestion and wrote down the affair as an accident.

Stephen Reed was the one man in the North of England who could have seriously improved the safety of the coalmines had he had the integrity to speak out at these Inquests, but he never did. Mathias Dunn on the other hand had by now got Reed's measure. Dunn contributed to a paper on the Burradon Explosion and tried to influence the Institute. With hindsight it is obvious that mines like Burradon were run on a knife-edge: there was no margin of error, any one or two of a dozen likely events could tip the balance into disaster. However, Dunn's efforts resulted in the Burradon Company building a larger ventilation furnace, and Burradon went on to become the oldest coalmine ever when it closed in 1975.

A fund for the widows was started and over£5,000 was collected throughout the North, the contributors' names being published in the Newcastle Journal. Lord Ravensworth gave £25. Joshua Bower, the owner of the Colliery gave £400. Stephen Reed gave £5.

Some say that the flapping of a butterfly's wings in China can eventually alter the course of European History. In other words, it is useless to speculate on what might have happened had some trivial event occurred or not occurred. Yet it is tempting enough to ask what *could* have happened if Stephen Reed had been a man of conscience. Certainly, there would have been fewer lives lost. But the poor illumination given by enforced safety-lamps, larger ventilation roadways, bigger furnaces, more safety rules, damages awarded now and then from culpable employers, would have meant dearer coal; less wealth and power would have been concentrated

The A and B Pits at West Moor (High Main Seam)

*Two pits sharing one shaft. The rectangular pillars
had their longest sides in the direction of easiest
hewing. The hatched areas show where the
pillars have been extracted "coming back brokens."
There were at least five staple pits, or underground
shafts, at Killingworth and one can be seen here
near the main shaft. The workings shown here are
only a small part of the Colliery.*

[The Coal Authority]

The C and D Pits at Killingworth Village (High Main Seam)

Incredibly, the shaft was sited on the boundary
of two landowners and this led to a law-suit.
Support has been left for Killingworth House but the
neighbouring land was worked to its boundary to provide
royalties for its owner. Note the shaft support pillars.
The workings south of the shaft may have stopped
short of working out the district because of flooding,
the coal dipping to the south. Until then the water would
have been removed in tubs by boy water-leaders. The pair
of horizontal roadways leading to West Moor can
be clearly seen.

[The Coal Authority]

Killingworth Village, 1858

(pictured opposite)

The well below Killingworth Hall would be easier for the villagers to reach than the Letch further down the path. The path, extreme lower right, also leads to a well. The area marked (190) is the pit pond. Directly below the High Pit shaft (189) is the by now disused waggonway where George Stephenson had built a self-acting incline, with full waggons going down hauling empties up on a single track with a by-pass in 1819. This coal was sent to the river as "East Killingworth" for a while. Note the allotments given to favoured workers. "Plumper's Row"* would most likely be so-called by the Ordnance Surveyor because the miners living there daily "plumped" directly on the end of a rope to their work, unlike their colleagues at West Moor, (and just about everywhere else!), who had cages with wood guides. Coal was not being wound at this time and the descending men would be ventilation furnace men. Perhaps the odd corf was drawn for fuel for the winder. The small building, upper left of the garden area (169) was the "Leper House", by tradition, and was ancient. The grounds of Killingworth House were surrounded by a high brick wall heated by internal flues for fruit growing. The church had not yet been built.

)n the other hand, "Plumper" was also a term used at that time for a voter, and it is just possible that some foremen, with rateable values of £10 per year, lived there.

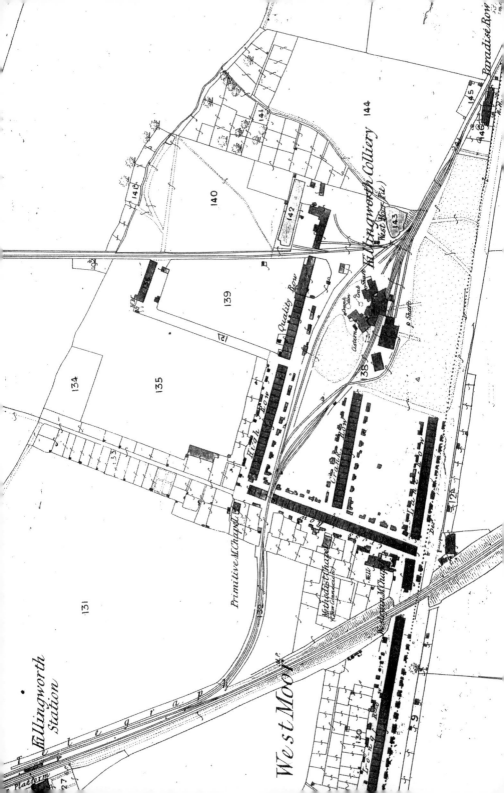

West Moor Village 1857

The "A" shaft is situated in line between the Crank, the old beam engine, and the Counterpoise Chain Well to the south. The shaft to the right most likely is a shallow opening to assist pumping and upcast ventilation, being connected by a drift to the main shaft somewhere above the High Main Seam. It is not shown on the High Main working plan and is therefore unlikely to have reached as far down. It was certainly not available for rescue when the main shaft collapsed in 1882.

Note the allotments and the three chapels. The small out-buildings are wash houses, not outside lavoratories, but they could have housed a chamber-pot; only after 1872 were dry privies built. Killingworth Station lies on the main railway line which cut straight through the village. Killingworth Village lies a mile away to the north east.

in the British aristocracy, and Imperial expansion might have been slightly restricted; the industrial revolution would have slowed perceptibly; England would have become comparatively less powerful in the world, and today everything would be slightly different. Reed possibly had more effect on world history than many a general. He, of course, probably never thought much about what he was doing, apart from consciously keeping on the side of established order, while the revolutionary tide flooding the Continent insistently flowed where it could find a footing in England, a tide which had swelled until 1848 and then began to ebb. Britain's prosperity was growing all the while, built on coal, some would argue upon the backs of the miners. Britain certainly became a very good place to be born in. It still is. But then, as now, it was better for some than for others. Anyone living in Britain now, owes a debt to the 19th Century coalminers for at least part of his present high standard of living.

1870

June 13th. A Monday. The West Moor School diary says that many boys were absent on account of the Miners' Picnic at Blyth. The Colliery would be closed for the day with a holiday without pay for the men. The Miners now had the Miners Mutual Confident Association. In 1944 James Bowman, General Secretary of the Northumberland Miners' Union, later Chairman of the National Coal Board, paid tribute to William Hogg, their Treasurer, when he retired aged 79. Hogg had been born at West Moor and could recall the West Moor branch of the Association marching out of the village with its banner, led by a German band, on its way to a mass meeting on Newcastle Town Moor. He also remembered visits paid to his childhood home by two men, one carrying a drapery pack and the other a stock of tea, who canvassed from door to door to earn a living, selling their goods: a way of life forced upon them by reason of their union activities, both being "sacrificed" workmen.

October 15th. George Mordew, 67, a Shifter was killed by an explosion of gun-powder. The black powder used was liable to spill, unseen, in the coal and was easily ignited accidentally; even at its best, compressed into "bobbin" form, the ignition of black power shots was somewhat haphazard, and even into the 1930's was causing similar deaths in other pits.

1871

October 19th. J. Morpeth, 56, a Shifter (daily-paid man) was killed by a fall of stone.

1872

New Coal Mine Regulations, growing awareness of possible accidents and their prevention, and good management by Mr. Crone, all began to have their effect. There were no deaths underground in that year, and until the pit closed only three more workmen were to die.

In October, 1872 the "Newcastle Weekly Chronicle" visited West Moor as one of "Our Colliery Villages". The reporter walked from Benton Station and found the village picturesque in the distance as he approached but dingy and smelly when he arrived. He said it was almost as bad as Seghill which the week before he had called the worst village in the County. Ash-heaps lay parallel to the rows, where chamber pots were emptied. There were no privies except at the schools (attended by 230 children). An old man told the reporter that he did remember a privy nearly twenty years since, but now "there was nowt for them but breaking through a hedge." The rows had almost flat-roofed "lean-tos" at the back, to increase the space inside each house. Many of the houses had four-poster beds, the uprights of which were said, half-jokingly, to keep up the sagging ceilings which leaked in rain. There were small separate out-houses which served as wash-house and pantry, and an open drain, here and there roofed with boards. All ash, refuse and excrement was thrown onto the ash-heaps.(1) There were large outside communal ovens for baking bread. All the gardens were well away from the houses, which therefore appeared bare and drab. Water had to be carried and many families had small casks fitted on to wheelbarrows to collect their water from a well.

In December, 1878 the "Newcastle Courant" paid a similar visit to both Killingworth and West Moor Villages. It mentioned the Killingworth Arms and The Plough and the castellated fronts of houses opposite the mine.

Children at Killingworth Village. Looking up towards the High Pit from the Church, 1893. The ners' cottages to the right became known as Plough Row after the Plough Inn on the same side, further up the hill [North Tyneside Libraries]

Crenellated Buildings, Killingworth Village 1930's.

[Norman Whitelock]

The High Pit was no longer drawing coal, but the winder was still there, together with a boiler embedded in brick. At West Moor, the doors and window-shutters were painted red, the door-steps were pipe-clayed and each house had a large water-butt. Drinking water was still wheeled from distant wells. The old crank-gin near the shaft was disused but still had its rope (hemp) attached, protected from the weather by a roof. Recent sanitary legislation had resulted in some change of appearance of the pit-rows. (Probably, the construction of dry-privies or "netties". There was no question of water-closets, local authorities long dissuaded their construction, there being no adequate drainage system.) The four-poster beds were again noted, as was the presence, here and there, of a rosewood-cased clock.

Both writers noted the presence at West Moor of working locomotives dating from George Stephenson's time: with four wheels, the cylinder on top of the boiler. By 1872 engines of this type were already kept as curiosities at the South Kensington Museum.

1874
January 13th. Joseph Dixon, 24, hewer, was killed by a fall of stone.

1877
April 14th. John Campbell, 34, hewer was killed in an unusual manner. A piece of stone fell from the roof and struck his sharpened pick, forcing it into his left side, cutting an artery, causing him to bleed to death.

Technical Developments

By this time, Mr. Crone was well established as General Manager of Killingworth, and was respected by his fellow engineers in the Institute for his technical ability. Among his innovations was the extraction of the pillars of coal left in the early days of Nicholas Wood, and of the highly inclined-areas of the High Main seam, until then considered too difficult to work. The seam was over 6 feet high, all clean top-quality coal, but was highly inclined to the 90 Fathom Dyke. He drove a level heading on the strike-line with well-supported headways to the full rise, every fourth

Killingworth House

otographed in 1898. The year the General Manager, Stephen Campbell Crone died.
[North Tyneside Libraries]

Stephen Campbell Crone.

The last Head Viewer at Killingworth
[Mr. J.R. Crone]

one being equipped with a "dilly" or self acting rope haulage, a hemp rope (later a chain) around a pulley allowing the fulls to pull the empties up, a lad at the top controlling the speed with an iron sprag in the spokes of the sheave. Horizontal bords were then driven either side, creating pillars 12 yards by 20 yards, which were then retreated horizontally with 6 yard places, the roof being allowed to collapse in a controlled manner. Sixteen pillars were extracted to each "dilly".

All working was now carried out according to the rules of the Coal Mines Regulations Act, 1872.

FOOTNOTE

(1) A late winter walk across any ploughed field in a mining district will discover fragments of broken crockery, thrown long ago on to an ash-heap in the nearest colliery village and brought by the farmer with the rest of the deposit to fertilise his crops. He took the ordure when it suited him, and if his field lay fallow, the ash-heaps grew bigger and smellier.

CHAPTER 5 - The End 5 April 1882

The end of the "slaughter-pit", when it came, was sudden, and not without a parting shot.

At six-o-clock in the morning, Andrew Carr, 55, Master Shifter, who more than thirty eight years before had been in trouble for throwing stones underground at a Deputy, and whose son Peter had been killed by a fall of stones 12 years ago, left his home at No. 9 Office Row, joining John Palmer, Deputy, also of 55 years, from No. 7, and set off to walk through rural Killingworth Village for the old High Pit, which served as the upcast shaft for what was now, in effect, West Moor Colliery. The green fields were closely cropped by Colliery horses, so that the corrugations left by ploughing during the Middle Ages could plainly be seen, stretching downhill in the direction of the Tyne.

Through the early morning mist, they would just be able to see beyond the River to the rising hills of Co. Durham. They passed the small "Lily Pond", a sump resulting from the medieval field drainage being disturbed by underground subsidence from coal-working. Ponds were a feature of the landscape. The refuse heap from the West Moor Pit had affected the natural surface drainage of the land, one source of the Ouseburn being cut off so that it could only deliver its water to a pond on the Colliery premises. But it was the subsidence, taking several decades to consolidate and settle, which caused the water to collect in hollows in the clay. The thick High Main Seam, when worked by pillar extraction, could produce dramatic surface effects; when Burradon, in the 20th Century, worked seams underneath the West Moor workings, the undulations of the fields increased and a giant pond, with reeds, and waves on a windy day, was formed, the hedges disappearing into its depths. When the New Town was built this latter pond, reduced in size, was walled in and became Killingworth Lake, a centre for water sports.

At the High Pit the two men were joined by John Flint, 58, Overman, two more Deputies, John Shields, 46, and William Fail, 52 and old John Wardle,

75, Master Wasteman. While awaiting the ascension of the kibble from the depths the men would be squatting "on their honkers" (q.v.), not needing to lean while so doing against a vertical support, quite accustomed to resting in this manner, after their walk across the fields. They would be wearing cloth caps and suits of checked fustian, home made or converted from old Sunday suits by their wives. Their trousers or "hoggers", were just over knee-length, easily removable over their boots, for them to work in shorts, and not liable to trail in mud or water. They would have hidden their matches and clay pipes under likely looking stones around the pit, in defence against the marauding of small boys, then on holiday from school and, at that time of year, not engaged in part-time farm work (nor beating for hares at Gosforth Park, this latter, seasonal occupation being a serious cause of truancy); they would however, have plugs of tobacco for chewing.

Killingworth Village then had a very peaceful, rural aspect. Apart from the pit cottages at Blue Row and Plough Row, there were three farms on its one main street. Then, as now, it was regarded as a desirable place to live and a few business and professional men commuted from there to Newcastle. The House, also on the main street, had had a splendid view over its own land to the south, being built in the 1700's, and had its own "follies" of castellated architecture and even a triumphal arch. The Hall, across the road from the house, was occupied by an affluent brewing family. The men must have enjoyed the novelty, normally reserved for Sunday morning walks from West Moor, of their bucolic surroundings.

Both Flint and Wardle also lived at Office Row, which originally had been named Quality Row, built next to the pit for the underofficials. Wardle was a fit man for his years, probably because he had avoided the debilitating effects of hewing for the latter part of his life. He told people he was three years younger than he really was. If he had been short of stature, as well as a bit more mathematical that the average pitman, those qualities would have given him natural aptitude for his occupation and assisted in his selection for the post. The Master Wasteman led an isolated life, wandering the, at one time, seventy miles of ventilated roadway in the mine, more than half in the return or "waste" where the travelling way was seam

Group of locomotive workmen
West Moor early 20th Century. The surface sheds were kept going long after the mine closed.

[Norman Whitelock]

Long Row, West Moor 1910
The gas was produced at the colliery yard.

[Norman Whitelock]

height, and where "creep" and falling stones had reduced the headroom still more. Even the main rollyways were often no higher than the first suitable parting in the strata, sufficient to accommodate a pony.

It would not be until 1954 that legislation was introduced to make the height of travelling roads 5 ft. 6 ins. and then only for the main shift of men to travel. At the older mines, as at adjoining Burradon until modern times, all the wastes conducting the return air were patrolled at three-monthly intervals by the Master Wasteman, whose initialled and dated chalkmarks were left in neat columns on handy props and boulders at every junction and change of direction on his travels. It was rare for any production-rushed official to venture into his domain, and possibly dangerous and certainly illegal, for any other workmen to follow him. When men were needed for repair work they would be provided by the Overman for his direction. A modern mine has designated intake and return airways but in the old days, where pillars of coal had been left without the bords being stopped off, it was often more efficient to use the greater cross-section of airway so provided as a general air-course, and, at the same time, keep the old wastes from filling with gas.

Near to the upcast shaft the several return airways joined up. Such a laby-rinth had to be patrolled and maintained and reported upon. Where necessary, measurements of the air-ways were taken, the speed of the flow measured by gunpowder smoke-cloud or, latterly, by anemometer, and the quantity of cubic feet of air flowing per minute computed and recorded in the office. Master Wastemen were usually proud of this expertise and, together with Horse Keepers, Mechanics, etc, emphasised their profes-sional status by their dress, wearing ties, even underground! In later, Nationalised, days, the post became Ventilation Officer. John Wardle, at 75, was the archetypal Master Wasteman and before the day was out would show his survival qualities.

At the top of the smoking, upcast furnace shaft, with its wooden headgear, one imagines the conversation: "Take a deep breath, Jack," could have chaffed Flint, the Overman as they together climbed into the waiting kibble, at ground or "hole" level, watched by the other four officials and two newly-arrived shifters, Edward Charlton, 40, and John Orton, 62, unmarried, and 16 year old Edward Harris, a pony-driver. Down the shaft they went, three or four at a time in the large bucket, slowly, through possibly 150°F of heat and rising acrid smoke from the ventilating furnace below, for there were no guides to their conveyance and any swaying could have caused accidental collision with the shaft wall which their "Geordie" lamps barely illuminated, less than four feet away, and then to the clean, cooler air below the furnace exit. The two shifters and the lad went together, and on alighting at the shaft bottom, most probably spoke to the furnaceman, John King, 40, who was pausing from shovelling coal on to the grate to act as onsetter.

"Never mind a bit smoke. There's plenty of fresh air for you after tomorrow. It's Easter," might have said King. Easter Monday was since 1871 a paid Bank Holiday, and Good Friday had always been taken. "Aye," would have said Charlton, "but less pay."

"Now you are a greedy beggar," probably joked King, "what about those poor devils up there with no work for three weeks? We are the lucky ones."

Two hundred and forty men and boys were employed at the colliery when it was fully working. However, since 15th March, coal work had been suspended, following a small fall of stone from the West Moor shaft walling, when an immediate inspection by Mr. Robert Wright, the Colliery Engineer, had discovered two places in the shaft where the old timbers had decayed: timbers which had repaired previous loose walling, possibly incurred during the explosion in 1806 when George Stephenson saw "wood, stones and trusses of hay" blown out of the pit mouth, or maybe when bratticed divisions in the shaft were reduced from four to two, or maybe when the rioters of 1849 threw tubs and other missiles down.

Efforts were being made to re-employ men at nearby Gosforth Colliery, which had the same certificated manager as Killingworth, but in the main the only men in employment were those above mentioned, engaged in the essential maintenance work for ventilation and drainage, and four shaft men per shift under the command of William Alder, Master Sinker, from Seaton Burn, removing the old timber and rebuilding the West Moor shaft in stone. The underground maintenance men had, at first, travelled to their work daily down the West Moor shaft, until 23rd March. Then, after the cages had been taken out, to be replaced by two working cradles with kibbles running between their suspension ropes to deliver materials to the shaft men, they went via the upcast shaft at Killingworth High Pit, a mile away.

By that morning of Wednesday, 5th April 1882, the work at West Moor was proceeding well and already 45 fathoms of walling had been completed. It seemed the mine was to have a new lease of life. Having braved the rigours of the Killingworth furnace shaft, the men were in good spirit as they made their way underground towards West Moor, backs bent, aided by the, at times, steep gradient, their way lit only by their feeble safety-lamps. Within half an hour they arrived at the main shaftway. The party then split up, one group going to the stables for ponies needed for their work, probably leading water after hand-bailing into tubs, where dips or "swalleys" (q.v.) in a roadway threatened to cut ventilation.

Pumps were operated normally by endless-rope from a haulage system driven by steam passing down the shaft through insulated pipes, and also by a rope run down the shaft, (the system also used at nearby Burradon). There would be difficulties when the shaftmen were working. It may also have been considered necessary to fire an inbye boiler to operate a pump. By the end of the century, electricity would greatly ease the pumping problem, and eliminate dangerous boiler-fires underground, but not yet.

Then John Palmer, Deputy, noticed that something was wrong and turned towards the shaft bottom. There was no air current! Gradually, by his poor light, he saw that the shaft bottom was blocked with broken timber and stones. At first he tried to clear a passage by removing some planks but soon realised that the shaft was choked with an immense quantity of

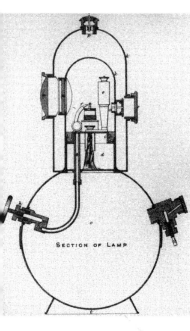

The Fleuss Lamp
The Fleuss Lamp taken to Killingworth High Pit rescue but not
used.

[Transactions N.E. Institute]

SECTION OF LAMP

The End of the "A" Pit
fter the winding engine was removed at West Moor. The supplementary pulley-wheels have also
gone. To the left is the ventilating chimney near the shaft mouth, now covered in rubble. The
chimney to the right was used to produce steam for underground use.

[Norman Whitelock]

rubble. Helpless, he turned back towards the stables, where the others, feeling the want of air, had gathered. Already, the partial vacuum, caused by the furnace at Killingworth, was drawing methane and carbon dioxide gas from the old waste, and lowering the percentage of oxygen in the air. It did not take old John Wardle, with personal recollections of at least five gas explosions at the pit, and an intimate knowledge of the ventilation of the mine, to tell them that there was only one way out and that to linger meant death! Edward Harris, the young pony driver, soon joined by the others, on the instructions of Flint, the Overman, was already piling choppy beside each of the forty or so ponies, chained to their stalls in the low, whitewashed, unlit gallery, 174 fathoms below the surface, to give them a chance of survival until they returned. Wardle possibly knew that at Harraton Colliery, after an explosion earlier in the century, a pony had been found alive on re-opening the mine seven weeks after the shafts had been sealed to deprive an underground coal fire of oxygen!

Orton, the 62 year old shifter from Benton Lane, West Moor, was of a nervous disposition and began to panic; he had respiratory problems and was already feeling the effects of lack of oxygen. Led by Palmer, with young Harris soon running ahead, the men set off along the tortuous route to the Killingworth shaft, this time climbing steeply as they went, leaving the stables in darkness forever, and the ponies, where they still remain to this day.(1)

A short time before, at the West Moor shaft, four shaftmen and the Master Sinker, Alder were standing on the cradle, contemplating their new foundation for another upward layer of walling, when suddenly there was the most thunderous echoing noise below them, followed by a rising cloud of dust. Hardly realising what had happened, but conscious of the precariousness of their situation, they quickly signalled to be returned to the surface, squeezed into the kibble or clinging to the rope with one leg through a loop in the ancient manner of riding from the pit, all five of them at once. The time was 6.30 a.m. Once safely at bank, Mr. Wright, the Colliery Engineer, was sent for; it was estimated that the fall had occurred at a depth of 70 fathoms.

At 7.30 a.m., with the noise of a cannon, a very large fall occurred and a messenger was dispatched to Killingworth House to seek the Agent, Mr. Crone, then aged 56, a widower, and his son, Edward W. Crone, 29, unmarried, the Certificated Manager of the Colliery. Two Deputies, Archie Muir and William Gilchrist, were sent to Killingworth High Pit to go down and fetch the men out.

The effect of the falling stone was cumulative; loose rock no longer held in place began to fall at intervals during the day. Each fall could be heard like thunder, two to three hundred yards away from the shaft, which no one was allowed to approach. Possibly, the pillars of coal left in to support the shaft had been inadequate, and the whole of the ground was unstable.

Meanwhile, back at Killingworth. Muir and Gilchrist had climbed into the kibble and been lowered down almost to the bottom, at 84 fathoms. The shortage of oxygen for the furnace meant that although some fire remained in the bars, carbon dioxide and deadly carbon monoxide fumes were being produced. On shouting to the surface to maintain contact, the two men found they could not be heard, and somewhat alarmed, they signalled to be returned to the surface. On their arrival there they collapsed, and had to be lifted out of the kibble. The situation now became very serious and an anxious crowd began to gather at the surface. The second editions of the local newspapers for that day held out little hope for the nine men and a boy trapped below ground: by 12 noon no communication had been heard from them, although the Killingworth shaft was equipped for signals by pull wire in both directions.

Mr. Crone Snr. took charge. Telegrams were sent to Seaham, where the Marquis of Londonderry kept for emergencies what amounted to an embryo Mines Rescue Service, equipped with the newly-invented Fleuss self-contained breathing apparatus and a huge, stretcher-borne lamp which provided its own oxygen. (The lamp would have burned under water. A spirit lamp burned lime (as for limelight at the theatre) and oxygen from a platinum jet played on the flame). Working downwards from the surface, a brattice was constructed, dividing the shaft into two parts, using tarred

The High Pit, Killingworth Village, date unknown but published 1881
The rectangular wall remained for eighty years after the Colliery closed. It existed in 1858. From at least 1844 this shaft was entirely upcast, ventilating the whole colliery but aided by another furnace at West Moor, probably until 1852. Coal-drawing in shippable quantities may have stopped at the High Pit about 1842 when cages were introduced at West Moor. The house, centre-background, is Killingworth Cottage.

[The Illustrated London News Picture Library]

canvas, and a fire lighted in a small drift at the surface. The convection air current produced was thus gradually advanced so that eventually it could clear the deadly fumes at the shaft bottom. The men so engaged were suddenly startled by the sound of pull-wire raps from underground. A kibble was sent down but came up empty. William Alder, the Master Sinker, and Ralph Gray, a shifter then descended, again to within 8 fathoms of the bottom, where their lamps were extinguished by foul air. From that point they shouted to the depths but received no reply from the men and were compelled by the fumes to ascend. Bratticing continued until 1.00 p.m, when, suddenly, there were five raps from below. The kibble was again lowered but again returned empty.

Three trial Stephenson lamps were then sent down and came back extinguished.(2)

The brattice slowly advanced downwards and by 6.00 p.m. the lamps came back up in the kibble still burning. It was nearly 10.00 p.m. when the Londonderry men arrived, superintended by Mr. C.R. Barrett, Manager of Seaham Colliery, with their cumbersome apparatus which was started to generate a supply of oxygen for the Fleuss breathers. Alder was sceptical and afterwards complained of the delay. Septimus Hedley, Assistant Underviewer, and Thomas Wetherall, Overman from Seaham, and Edward Howard, Master Shifter, from Killingworth, who knew the route, then descended, wearing breathing masks but carrying a Stephenson lamp, and immediately returned, saying that the "Geordie" was burning satisfactorily, but that the Killingworth volunteer's mask had become disarranged.(3)

Their difficulty with the apparatus, particularly with the cumbersome lamp, was too much for Alder, who perhaps felt partly responsible for the shaft collapse which had caused the situation, and he, the young Manager, Edward W. Crone, William Lawson and Ralph Gray went down in the kibble and this time reached the bottom. Alder and Lawson penetrated inbye for 55 yards and found the men all alive, but only King, the furnaceman, and old Wardle conscious. King would not enter the kibble until the rest, whom he described as all "much worse than him," were saved.

95

The three Seaham men then descended with their apparatus which was found of great use in giving each of the exhausted men a "whiff" of oxygen, which revived them much.

The first man brought to bank was Andrew Carr, then old man Wardle, then Shields and the boy Harris, then Flint, Fail, Orton, Palmer, Charlton, and finally, at 11.55 p.m. King.

As each man was lifted from the kibble he was attended to by Dr. Joell, the Colliery Doctor who lived in Killingworth Village, with stimulants and then, in one of the adjacent old cottages, by Dr. Buttercase and three students from the Newcastle College of Medicine, Messrs. Blain, Buxton and Britton-Parkinson. The medical men were optimistic that all would recover, although Orton appeared to be very ill. The next morning some of them were able to walk home but others had to be conveyed. Orton had worsened and died at home at 9.00 p.m. that evening, from "nervous prostration"; it may be, one now suspects, that cumulative carbon monoxide poisoning contributed to his death. Palmer related afterwards that Orton had been the most despondent of their chances of survival; he himself remembered nothing between waking up in the old cottage and sending up the five raps at 1.00 p.m. Young Harris had sent the earlier signal, having run ahead of the party. It was just as well that none had been able to enter the kibble, as they would almost certainly have died as it passed the furnace drift. At the Inquest on Orton, at the Killingworth Arms, evidence was also given by J. Stoves, a fitter and Thos. Hepple, shaftsman. Later, in his Report, the Mines Inspector described Orton as "feeble minded", but this was probably unfair: a six hundred feet climb in very trying circumstances, for an unfit, 62 year-old, may have been just too much.

The falling of the stone at West Moor became a wonder of the neighbour-hood and could still be heard two weeks later, by which time the winding engine house began to subside, and the whole of the workforce was laid idle for an indefinite period, although eventually most of the men were employed at other collieries owned by John Bowes & Partners, princi-pally at Dinnington, to where the men were carried daily by a Colliery locomotive towing old passenger carriages until the 1930's; Gosforth closed at the same time as Killingworth, being interdependent. After a month or so the shaft was filled in with rubble from the extensive pit-heap, with the intention of digging it out and re-walling by stages. However, better judgement soon prevailed, the mine was abandoned and much of its remaining coal worked from Burradon, which by 1914 had over 1,400 men employed underground.(4) The shaft made its final, earth-shaking protest in 1944 when a large subsidence occurred as a hawker's horse and cart was passing over the unmarked spot, although both were eventually rescued with some difficulty.

Somewhere, beneath a modern industrial estate, no-one knows exactly where, the West Moor "A" shaft still stands, filled with stone, and down below, in flooded galleries, the bones of ponies and the ghosts of men lie forever.

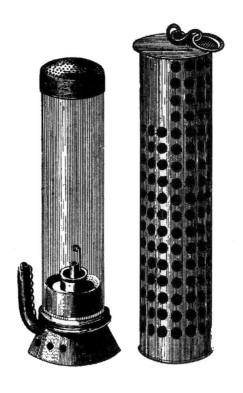

The "Geordie" Lamp
Invented by George Stephenson
and developed by Nicholas Wood
who replaced the perforated plate by gauze
[N.E. Institute]

FOOTNOTES

(1) Long burning flame safety-lamps equipped with colliery-made, enlarged oil reservoirs existed but were reserved for shaft-bottoms and possibly other places where several men worked together. Stables were usually kept in darkness, even until modern times, except when men were in attendance. The author remembers acting as a temporary horse-keeper during the annual pit-holidays at the Rising Sun Colliery, Wallsend, in 1951 and 1952: the then electric mains lighting was switched on daily only to clean out the stalls. (In 1952 the one-week holiday was doubled and the thirty-three ponies in the Main Coal Seam produced sixteen tubs of manure over that time. There were other stables elsewhere in the mine.

(2) By that date the Stephenson lamp was outmoded, although these were slightly modified versions of the original design. Its continued employment at West Moor had been remarked upon with some pride by Nicholas Wood twenty years earlier earlier at an Institute meeting; he had been involved in its invention, but this approval was mainly due to its ability to safely withstand draughts up to 13 feet/second, higher than the Davy or Clanny lamps, and also to its property of extinguishing itself when gas burned inside the glass. Unfortunately, this meant it was possibly to mistake black damp for fire damp. It did not give as good a light as the Clanny which had clear glass around the flame. Another danger associated with the modified "Geordie" or Stephenson lamp was that the original perforated outer plate had been replaced by gauze of the Davy type and the inner glass cylinder shortened: while testing for gas a deputy could grip this glass through the gauze after turning the lamp on its side to work the glass up away from the flame, thus allowing the atmosphere to quickly reach it, giving a better indication. However, this meant that the lamp became a Davy lamp of large diameter, dangerous to explosive air-flow and also, as came to be realised at about this time, that unconsumed gas filled the raised glass tube, and the explosion resulting from its ignition at the reduced wick flame undoubtedly did, on some occasions, carry the flame through the gauze, and caused loss of life without leaving any trace to indicate the origin of the explosion. Within a decade of the West Moor shaft collapse,

experiments by Marsaut had led to the general adoption of the bonneted and double-gauze, windowed Clanny lamp still used today for gas indication. The old, plain gauze Davy lamp, however, continued to be used for many years at some collieries by officials, for gas testing rather than for illumination, which, of course, was the main purpose of the workmen's lamps.

(3) The Fleuss apparatus used compressed oxygen and was based on a self-contained diving apparatus of the time; it had a face mask which was uncomfortable to wear and a contemporary "Colliery Manager's Handbook" noted that "even without a beard a man is uncomfortable". Training is required to safely use self-contained breathing apparatus. The author remembers wearing for the first time, in training, the not dissimilar Brown-Mills liquid air apparatus in 1952 and having what proved to be an irresistible desire to tear the thing out of his mouth; two rescue men died during rescue work at the Easington disaster in 1950 wearing such equipment. In 1882 the Fleuss Apparatus, with its lamp, was advanced for its time, and immediately after the West Moor shaft collapse negotiations were commenced with the proprietors of the patents to supply a complete and large set to the Northumberland Coal Trade Association, to be kept in Newcastle, but from the conditions and restrictions named and absolute refusal to sell, the negotiations fell through.

(4) Including the Lizzie Pit at Weetsdale.

Miners And Families at "Old Pit", Killingworth Village 1905

The little girl at the front, aged three, is Jane Thompson who
was the last person killed by the High Pit; she drowned in the
pit pond the following year.
There was a great gap between owners and workers. Nicholas
Wood, who built this house, died at his residence at Sussex
Gardens, Hyde Park in 1865 while still maintaining Hetton
Hall in Co. Durham.

[Thompson Archive]

101

GLOSSARY

ACCIDENT

A misfortune which was not foreseen. Or so it tends to be defined nowadays. According to modern industrial thinking, an accident at work which "was waiting to happen" is negligence on management's part; indeed any accident, even when an employee has behaved in a near-suicidal manner to cause it, will usually result in financial compensation being paid by the employer for his lack of foresight. Very rarely does a court decide that an employee is entirely to blame. Many jobs are inherently dangerous, and it is up to the employee to look out for himself to a certain extent, but management is obliged to minimize the risks, and certainly to pay financial recompense after injury or death.

Things were not always so, and the change has been gradual. The numbers of fatalities in coal-mining has slowly declined over the years, ever greater efforts being needed as each successive major cause of accident has been eliminated. Virtually all mining safety legislation is the result of particular accidents in the past. The more obvious regulations being the first to be enacted, to counter iniquitous and horrendous situations, the later laws gradually safeguarding against the rarer and more obscure causes of injury and death.

Looking through the vast catalogue of 19th century fatalities in the accumulated Mines Inspectors' Reports, numerically four major causes of death stand out:

1. The use of naked lights. The tallow pit candle of 30 per pound gave only half the light of a sperm candle, or a modern one of wax, but it was still at least four times as effective as the first safety lamps, and in the case of the Stephenson lamp used at Killingworth, about seven times so, and consequently there was always pressure from both men and management to use naked lights were the situation seemed safe enough. Unfortunately, conditions could change or workmen could enter more dangerous parts of the mine still carrying their candles; many deaths were caused this way.

2. The lack of systematic timbering (q.v.). So long as supports were placed only where considered necessary by the Deputy, falls of stone occurred killing men. That the Deputy, not the workman, was responsible for the physical setting of the props made things even worse. A letter to the "Colliery Guardian" in 1882 called for the men to set their own props, not further apart than a distance specified by the manager, a system which the writer said was working satisfactorily in Staffordshire. But it took many more years before systematic supports become law.

3. Poor illumination below ground. Fatal accidents happened time and time again because men could not see properly. Men and boys fell down shafts, stepped into holes, tripped over haulage ropes, failed to timber dangerous roof-breaks, died in a hundred ways because their vision was limited to a few feet around their stationary lamp. They were specifically forbidden to carry their lamps with them when working (to reduce chances of gas ignition) by the General Rules of Killingworth Colliery of 1855 and 1869, so a putter would hang his lamp at the landing (q.v.) and ride in the dark on the limbers (q.v.) to the light of each hewer he served. This seems unbelievable nowadays, and he probably often disobeyed the rules. A deputy setting a prop might hold the handle of his quite heavy lamp in his teeth so as to light his work.

4. Lack of adequate medical care by present-day standards. Men would be carried home on carts or even walk there, only to die a day or two later, presumably from after-shock, the affects of burns, cumulative gas poisoning, loss of blood, internal bleeding etc.

Thus, nowadays it might be fair to say "there is no such thing as an accident, it must be caused", but at the same time it might also be said that "a pit is only absolutely safe when it is closed". A cynic might add the corollary that the process of making our pits safer has hastened their closure, but British Coal experts maintain that the present day cost of an accident, in fact, exceeds the cost of its prevention. Unfortunately this was not the case in the 19th Century.

AGENT

The General Manager of a mine or group of mines. An Agent had, in effect, control over the Certificated Colliery Manager who was legally responsible for the immediate working of the mine. He was the Colliery Company's Agent. He might also have acted as FITTER (q.v.). Coal Owners (q.v.) also had their Agents, who were independent mining engineers.

AXE

Pronounce "aixe" as in "takes". A long-handled axe with a notch in the blade for nail withdrawing, standard issue for Stonemen, who usually blunted the blade by using it to dig holes; Drawers, on the other hand kept their axes razor-sharp for cutting through difficult props, and well-hidden from Stonemen.

BAIT

Mid-shift refreshment. Bait was not regarded as a meal, rather a snack and no official time was allowed. It was taken during a suitable delay when there would be no cost to the miner from loss of output. Many a bait was eaten while walking outbye or even while turning a coal drill with one hand! Invariably, it was two or three slices of bread and jam, with no butter, which was reckoned to cause heartburn, washed down with water from a tin bottle. The bait was wrapped in newspaper and hung with string or, latterly, with used electric detonator wire, from a roof support, to stop the mice getting it, until required. (This system was proof against mice but not always against ponies!) Coal dust was kept from the bread while eating, by holding it in a piece of torn wrapping.

The ethos of bait in a coal mine contrasts with metal mining: the Pasty with one half for meat and the other for sweetened apple, enjoyed by Cornish tin-miners, or the iron-ore men in Cumberland actually slowly cooking a meal on a candle during the first half of a shift. A modern Australian metal mine will have "crib" room especially constructed with tables and chairs to which the miners are summoned by buzzer at the set time, but the "bait", of casual esteem and minimal proportion, lived on to modern times in Northern coal mines. Probably the difference had its origins in the short six hour hewing shift keeping hunger from building up too much, the comparatively cramped working conditions not allowing for any indigestion without pain and loss of output, and the long tradition of spending as little money as possible on work, while contriving to live as extravagantly as wages allowed at home. South of Northumberland and Durham, bait is called SNAP.

BANK
The surface. (In Cornish tin mines called GRASS.)

BAROMETER
A mercury barometer is nowadays posted by law at the top of the shaft so that any drop in atmospheric pressure is brought to the notice of officials. A lower barometer than usual will allow gas held in old workings to expand and spill out into ventilated roadways. In 1860, the Burradon Under-manager had one barometer which he kept at home to be shared with Coxlodge!

BRATTICE
A division in a shaft or roadway, originally of wood boarding, but for a hundred years until the mid 1950's of stiff, tarred canvas, to direct the ventilation. Plastic sheeting has since been used for this job. A modern brattice is usually to stop the air-flow completely as a temporary door. Auxiliary fans and ducting are now being used to ventilate headings.

CAGE
A wrought-iron frame, usually with two decks and a protective roof, suspended by chains attached to the winding rope. Usually two tubs were carried on each deck. The cage ran between two wooden guides so that small clearances allowed high winding speeds without fear of collision. Later cages were fitted with automatic detaching hooks to prevent danger from overwinds, the cage being held by retaining catches in the headgear. Other developments were rope and steel rail guides in the shaft and axle catches on the floor of the cage decks, which could allow tubs pushed in

to the cage to push out the empties with the aid of gravity, and be themselves retained in place for hoisting. In the late 19th Century, however, the empties were manoeuvred over flat-sheets by a half dozen lads for distribution inbye. The full tubs were eased into the cage on the other side of the shaft by the Onsetter and his assistant. The tubs were held in the cage by a swinging drop-bar.

CAUNCHING
The taking of stone in a roadway to provide sufficient height for horses or machinery. Men could travel in any height. Usually, in Bord and Pillar mining, a bottom caunch was taken for tub height if the seam was low enough to need it, the stones being built into packs on either side of the travelling way, which greatly helped to support the roof.

CAVIL
Pronounced variously CABLE or KYEBLE: the process of drawing lots for a working place, and also, the thus allocated place itself. The dictionary meaning of the word implies argument or objection and it must have been that the democratic drawing of lots eventually replaced the previous contention. In the Northern Coalfield, men were allocated to workplaces half-yearly and then quarterly, so that all had a fair chance of the better earnings available. It was also common practice for men underground to re-allocate different tasks among themselves by "Cavilling": usually chalk marks on a plank to be rubbed by the participants, the other side then revealing which work they were to do. Each colliery had long and complicated Cavilling Rules, written down in detail by the management and meticulously observed by the men. Both individuals and teams looked ahead to their next quarter's Cavil with mixed hope and dread. Sometimes a man might even be Cavilled to another pit, one nearby belonging to the same company, and he would have to carry or wheel his tools to the other shaft. The last Pay Friday (men were paid fortnightly) in the quarter would be "Shifting Gear Day" and work was finished early so that picks, shovels, etc. could be taken to the new Cavil ready for the next shift; this time might also be used for an eager inspection of the new place.

CHALDRON

Victorian term for a surface coalwaggon. During the life of Killingworth Colliery the weight of coal carried was increased from 2 tons to 4 tons. Early in the 19th. Century a chaldron (or cauldron) meant 53 cwt. A London chaldron was 25 cwt.

CHALDRON BOTTOMS

Known by various other names, mostly now corrupted and impolite, these are fossilised tree stumps left in the roof of a seam, and they killed many men, dropping out of their sockets, between the roof supports, when least expected. John Todd, 24, hewer, died, leaving a wife and two children, in 1886 after moving to Dinnington Colliery following the closure of Killingworth, killed by a Chaldron Bottom which fell twenty minutes after his working place was visited by the Manager and the Mines Inspector.

CLANNY

Now pronounced GLENNY. The term used nowadays for the Workman's Flame Safety Lamp, as distinct from the stainless-steel bonneted Deputy's relighter lamp. Dr. Clanny of Sunderland in 1815 produced the first-ever safety lamp which was an impractical affair, insulated from the atmosphere by water, its air supply provided by a small pair of bellows operated by an attendant boy. By the late 1840's a manufacturer was advertising a 'Clanny' lamp which was simply a Davy lamp, but with a glass tube replacing the gauze at the position of the flame, to give a better light. This latter lamp was not universally adopted because of the danger of the glass breaking. In the 1880's, after experiments by Marsaut and developments in glass technology, the Clanny lamp, now double-gauzed, bonneted and locked, became standard issue to workmen. It was produced by several manufacturers. It could be turned up fully without danger, the bigger flame simply drawing in more cooling air. This is not to say that it compared in illuminating power with a modern electric cap-lamp. Users were liable to contract nystagmus by working with their eye too close to its flame; to function safely it had to be hung, not stood on its base when it might not be truly vertical; if jerked it could go out and would then have to be taken outbye to a relighting station. The only way for a man alone without a light to find his way to the shaft bottom was to crawl, feeling the rails as he went, perhaps for a couple of miles or so.

CLEAT
The natural vertical fracture of the coal which usually lay in the same direction over large areas, almost invariably over the whole Northern Coalfield from North West to South East. A bord would approach these rectangles flat-side-on, which meant that a swung pick blade would naturally find the break and peel off the lump. A heading, or "headways" place, of necessity at right-angles to the bord, would approach the rectangles of coal end-on and thus each blow would have to smash its way through solid coal; the output from an undercutting hewer's shift in such a place might be two tubs of very small and very hard-won coal.

In addition to the effect of the cleat, the hewing might be eased or hindered by the presence of "slips". Fractures with slippery surfaces in the coal produced by ground movements, they lay at an angle to the seam floor, usually persisting over about the area of a pit district. If the slope of the slip was upwards, away from the hewer, there was the danger that the upper wedge of coal would suddenly slide out without warning; in a thick seam this could kill a man. If the slope was the other way, however, the hewer had a ready weak break to attack. Because of these geological variations it was the practice in Northumberland and Durham to have quarterly "cavils" to allot the good or bad places fairly.

COAL OWNER
Invariably, the lessees of the coal to be worked were referred to by the workforce, the public, the press and even themselves as Coal Owners, whereas they were, in fact, owners of the Colliery only. The actual coal belonged to the landowner(1), who let the working of it on a long lease and who received royalties, or annual payments depending on the output achieved, usually with a cumulative fixed minimum charge written into the contract. A land owner also received payment for wayleave both above and below ground. Several landowners, often including the Church of England, might be involved in one mine and they would be proportionally

paid, annually, according to the districts being worked. The Killingworth royalties included Balliol College, Longbenton, Glebe, and Bewick and Craster, as well as that of the Earl of Carlisle whose Agent, Mathew Liddell of Benton Grange, reported to him regularly over forty years(2), closely looking after his interests. In 1845 Liddell commented unfavourably on the raising of water in tubs, suggesting that an influx of water could close the pit, recommending his Lordship not to renew the lease when it fell due. Independent consultant mining engineers such as Liddell, in effect, provided another layer of top management, whose views would have been taken into account by the local colliery directors when major decisions were taken; their real function was to ensure every last penny for their aristocratic employers, the real coal owners, who, in this respect at least, were the Idle Rich. It must be said, however, that the Seventh Earl of Carlisle introduced in 1850 a Bill into the House of Lords calling for a Mines Inspectorate, possibly prompted by his lessee at Killingworth, Nicholas Wood, and by the rising agitation in the Commons following petitions by workmen. The Bill was passed with little opposition although the Marquis of Londonderry, another Coal Owner whose family's capacity for causing misery among the miners was legendary, spoke against it, making vain threats of non-co-operation and lawsuits, should Inspections become troublesome. This same Marquis was alleged by Ellen Wilkinson, MP, writing in 1939, to have enjoyed sitting in his carriage to personally supervise evictions in the 1840's. He wrote to the "Mining Journal" (10.8.1850), disparaging Lord Carlisle, who "knows nothing of the working of coal-mines" and referred to "the real demands and mischief of this Bill under the influence of the political economists, philanthropists and ultra-humanity conceptions and feelings." He states that "as principal owner in Durham" he will "no longer support paupers caused by unavoidable accidents." He goes on to say:

"I will submit to all the penalties of the Bill rather than yield to the monstrous injustice of allowing a Government Inspector to survey with his people and take plans of my underground and old workings; nor will I yield to any charges and costs for maps and plans of my extensive coalfields at the ad-libitum direction of the paid Inspector, who will like no better business than placing the trade in absolute confusion."

There is no doubt that in a sense the real Coal Owners, the Dukes, Marquises, Earls and Lords deserved the execration cast at them by generations of coalminers: they appeared parasites of the first order, battening on the labours of the working class with rapacious arrogance; they lived in luxury in their castles, effortlessly depending upon the work of others. Of course, from their point of view, the coal belonged to them and they were selling it to their lessees. Although royalties fell in general over the duration of the Nineteenth Century, the owners did in fact get the "sixpence a ton" much alleged by Trade Union propagandists then and since. In the 1840's the Earl of Carlisle, owner of part of the Killingworth coal, received 27s. 6d. for each "Ten"(3) of 49 tons, in addition to 2s. 6d. for every Ten produced at adjoining Burradon by reason of an underground watercourse driven to Killingworth providing drainage. In 1843, the Sixth Earl, living at the architecturally splendid Castle Howard, in Yorkshire, received £1,337-3s-11d from Killingworth. The Seventh Earl, who succeeded to the earldom in 1848, was a prominent politician of the day who, after originating the Mines Inspectorate, became Lord Lieutenant of Ireland; he also published poems, as had his grandfather the Fifth Earl. Later in the Century another Marquis of Londonderry, presumably more highly thought of than his Inspector-opposing predecessor, became Lord Lieutenant of Ireland and after a spell as Postmaster-General was admitted to the Cabinet in 1899.

These latter aristocrats, although Rich, can hardly have been said to be Idle; however, many a genteel spinster cousin or niece lived frugally, or even very adequately, from a settlement or portion from a coal property. The Ecclesiastical Commissioners were among the biggest coalowners, making the Church of England immensely rich, providing good livings for the Bishops and their hierarchy, rather less for Vicars and their curates; their properties were widely spread and thus more certain of possessing profitably-worked coal. Similarly, Colleges, endowed by past benefactors with agricultural parcels of land suddenly transformed into much greater than hitherto providers of income, became rich; cellars of claret and good

Port, and Feasts for the Fellows at High Table, came from the ingenuity of the colliery engineers and the sweat of the miners.

The Lessees or Colliery Owners, however, were Victorian entrepreneurs: hard taskmasters with all the faults and virtues of their kind, some rash gamblers but others men of vision and intelligence, all made ambitious by the then very obvious rewards of success, their parsimony to their employees explained by their worries over the viability of their investment. In this latter respect it might be argued, that since the High Main of the Tyne Basin was worked out, few mines of the Northern Coalfield could provide fair wages and conditions to employees and still make a profit. Certainly, after Nationalisation, when wages and safety improved, mines began to close. Intense mechanisation, to improve productivity, failed to halt the closures, and after the post-war coal shortage crises, when the cry was "Coal at any price!", the number of pits decreased with ever-increasing rapidity as markets fell away; the coal was not worth the price it cost to mine without Government subsidies and controlled imports. That is unless another very thick seam could be found, as at Ellington under the sea bed, and a few other places.

CORF
A hazelwood basket carrying up to 6 cwt of coal, transported underground on a bogey or filled at the shaft bottom from smaller containers and hoisted up the shaft, often in threes, one below the other. A corf could brush the side of the shaft or a descending empty corve without great damage. Cage winding replaced corves in the 1840's. Shilbottle was probably the last Northumbrian pit to use corves, up to the 1860's.

CRACKET
A stool used by a hewer (q.v.). Three pieces of wood, usually with a carrying hole in the seat. The uprights might only be about three inches high and the seat usually sloped. In effect it often served instead of knee-pads, which were unknown. Apart from sitting or kneeling on it, the hewer might use it to rest his shoulder while reaching with his pick, deep into the undercut. The brothers Thompson had a fist-fight underground, when Bill

hooked Jim's cracket with his pick and pulled it out from under him, for a joke to amuse the Deputy and the Putter.

CRADLE
A platform suspended in a shaft by one or two ropes for carrying out walling work. Where two ropes are used they can provide guides for a kibble (q.v.). Single-rope suspension of a cradle is now obsolete and when used was the cause of at least one serious accident: the chains linking the rope to the hinged, two-part platform, not having been all attached, resulting in its consequent collapse.

CREEP
When the roof pressure was too great for the size of pillar remaining after partial extraction of a seam, the soft underfloor would rise, pushed up by the descending pillars, like a giant unstoppable steamed pudding, to the roof, preventing any further mining in the district. Various remedies were tried, never successfully, until it was found that almost total extraction could be achieved by collapsing the roof, the broken stone then occupying a greater volume and filling the space between the floor and the pressure-arch above the seam, the weight of the superintendent strata being diverted to the solid coal on either side of the working face, often beneficially softening it up prior to the further advance. Creep was important at Killingworth. (See narrative text.)

DEPUTY
Deputy-overman who would look after a particular "district" or "flat". The Deputy would supervise the work, and test for gas and general safety. During most of the 19th Century, a Deputy was the sole arbiter of what roof supports were needed at a working place and had to set them himself. He might also be charged with keeping the men supplied with sharp tools and with advancing the way (rails) when required.

By the time Killingworth Colliery closed in 1882, a Deputy might be timbering for up to 10 or even 14 men. There were calls in the technical press from mining engineers for men to set their own props, not further apart

112

than a distance specified by the Manager of the colliery. Apparently this system was working satisfactorily in Staffordshire. Eventually, systematic timbering was introduced in mines, but never at Killingworth.

Nowadays, by law, a Deputy is legally in charge of the operations carried out in his designated District, and the safety of the men in it, and must not be required to do any manual labour which might interfere with these two priorities.

DRIFT
A tunnel driven either level or at an angle across the strata.

DRIVER
A boy's job, collecting tubs from putters, forming sets, and delivering them to a "landing" where large sets of 40 or 60 tubs would be shunted and collected for rope-haulage outbye. Drivers were killed from time to time by being crushed between tubs when uncoupling them, and by not keeping their heads down when riding the limbers.

ENGINEER-COLLIERY
A man of some consequence, in charge of all mechanical engineering at the mine. Nowadays, the title "Engineer" has been devalued and is applied to mechanics, dishwasher repairmen and so on, but in Victorian times it was not awarded lightly. In the early 19th Century the Colliery Engineer would design and build machine tools such as lathes, when there was little collective knowledge to draw upon, would build steam pumping engines and even locomotives, and would produce detailed schemes for rope haulage arrangements, etc. to meet the general outlines proposed by the Mining Engineers (q.v.)

He would, however, be of the people, with no pretence to gentility, live in the village, next to the pit so as to be always on call, and would in later years, together with his immediate assistants, develop a running feud with the mining men as to relative prestige. (Rather similar to that which runs in the Merchant Navy between Engineers and Officers.)

ENGINEER-MINING

A title of prestige originally reserved for persons above local management level, such as mining entrepreneurs' agents (q.v.), well known figures in the then mining world. By 1850 the term was being used by Viewers to describe themselves, and in the present day is current for young graduate officials prior to their reaching managerial status.

EXPLOSIONS

See narrative text for details of individual colliery explosions. The initial cause of virtually all mine explosions was an ignition of firedamp (CH4). At Weetslade in 1951 there was an explosion of hydrogen (H_2) produced from electric locomotive batteries, and there have been claims of direct coal dust explosions, caused in the absence of methane, by blown-out shots and also by adiabatic compression of the atmosphere due to large roof-falls, but these are rarities. Firedamp, or methane, burns in air to produce carbon dioxide and water: $CH_4 + 2O_2 = CO_2 + 2H_2O$.

Thus, apart from burning the unfortunate individual near it, an ignition could be relatively harmless and was often so regarded. However, when air was in short supply, as was often the case in coalmining:- $6CH_4 + 11O_2 = 4CO_2 + 12H_2O + 2CO$. CO or carbon monoxide is a deadly poison and was usually responsible for most deaths after a firelamp explosion. Also note the presence in the products of the explosion of a large amount of water vapour; this would condense causing a partial vacuum which might draw a further supply of methane and air into the place to be rekindled by any still-burning material. A second explosion was a common occurrence.

However, the worst effect of a methane explosion could be the lifting up of inflammable coal dust which could then explode, with much greater violence over a much larger distance.

$C + O_2 = CO_2$ and $2C + O_2 = 2CO$

Again, the deadly carbon monoxide is produced in a confined space.

The eventual answer to the coal dust menace was the scattering of non-flammable stone-dust to dilute the coal dust lining roadways, and the building of stone-dust barriers (stone dust heaped on easily-dislodged suspended boards) to isolate any coal dust explosion to a limited area. It is stone-dust that makes a pit a filthy place.

During Killingworth's day, however, the roadways were not stone-dusted and the seam showing along the travelling ways would have shown jet-black; likewise the men emerging from the pit-shaft would have had jet-black faces (neither were there water sprays in those days) rather than the dirty, smudgy faces of today's miners. In early days some men perforce remained unwashed, and their bodies took on a sort of black shine; instances are reported of early miners being dressed up in comparative fashionable finery, away from the mine, showing black faces. This takes some understanding: at a newly-opened mine water might be unavailable and at the best had to be carried a long way; it was a familiar situation at the pit-village; but neither was the black "dirty", it was simply fine coal. Well into the twentieth century, miners often deliberately left a black patch on their backs, as this was considered to help them retain their strength which washing might have drawn away; it could be that in some cases this was a ready excuse to explain away a wife unwilling to rub her husband's back, where he could not reach, while he sat in the tin bath!

EXPLOSIVES

The only explosive ever used at Killingworth Colliery would be gunpowder, or "black powder" as it was known. Gun cotton, invented in 1845, would be too expensive, and nitroglycerine too dangerous. After Nobel stabilised the latter to make Dynamite in 1866 it might just have been employed in tunnelling through hard rock, but its shattering effect was such as to prohibit its use in coal, where a devalued product would have resulted.

Miners paid for what powder they used and it was usually supplied in "bobbin" form, the safety fuse, which burned at 30 seconds per foot, being stuck into the hole of the bobbin, no detonator being needed. Short pre-cut lengths were called "squibs". In the very early days of Killingworth, before the invention of safety fuse, goose quills were filled with loose gunpowder with haphazard, sometimes fatal, results. A contemporary objection to blackpowder was that, being black, it could be lost in the coal and then accidentally ignited; a greater, real danger was its fiery propensity to ignite firedamp. Generally, management preferred no use of powder at all, if possible. The term "powder" is still used for explosives underground although they are all now nitro-based high explosives, made non-inflammatory by the addition of cooling salts, and need detonators to work.

FITTER
An agent for a colliery, or several collieries, with an office on the Quayside, who arranged the sale and shipping of the coal. (Not to be confused with the mechanic of the same name.)

FURNACE
Situated at the bottom of the upcast shaft the furnace produced the convection current to ventilate the mine. Furnaces were legislated upon and prospective Colliery Managers did involved calculations concerning their design and operation, for their Certificates. The obvious undesirability of fires in potentially incendive atmospheres led to efforts to supplant them: steam jets, which proved inefficient, were widely tried and even a gigantic air-pump was devised and promoted by advertisement. Large steam driven fans began to be used in the 1880's but furnaces persisted here and there until the 20th century. A large furnace might consume four tons of best coal per day. It had to be tended continuously, raked to burn always at its greatest efficiency. Early pits used small coal and let the fire go out at week-ends, often with disastrous results.

GAS TESTING
During the first half of the 19th century a candle was often used for gas testing. The Deputy cupped one hand around the flame to observe the halo, ready to instantly lower the candle at the first indication of gas. He might then require the men to use their "Geordie" lamps only, forbidding naked lights. The dangers in this practice were twofold: the tester himself

116

was liable to cause an explosion or, at least, an ignition, (older Deputies of the time were regarded as lucky and skilful if they had never been burned) and, secondly, the erroneous assumption that it was safe to work with a safety lamp in an incendive atmosphere could lead to an ignition of gas by overheating of the gauze, especially in any draught. Such a lamp gave a poor light and a hewer would turn his flame up to the point where the gauze became red hot. Still, many old officials swore by the candle as the best indicator of gas: a gas cap on a candle might be five inches high. Of course, the real trouble was that early safety lamps just did not give a good light and both men and management preferred to use candles unless conditions enforced the use of lamps. There was also the false argument that if gunpowder was to be used, then naked lights, being less dangerous than gunpowder, could safely be used. Nicholas Wood was a proponent of this view, and at Killingworth candles were used in the "whole" working (where gunpowder was necessary) and lamps in the supposedly more dangerous pillar extraction (which did not need blasting).

Workmen's safety-lamps, primarily for illumination were, and still are, used as indicators of gas rather than measurers of percentages. A Deputy now uses a relightable spirit lamp which produces a blue gas "cap" at $1\frac{1}{4}\%$ and a halo in the shape of an equilateral triangle at $2\frac{1}{2}\%$.

Shotfiring is stopped if there is any indication of gas at all and workmen must be withdrawn if there is 2 %. The explosive limits of methane are 5% to 13%. Some slight sensation of a methane ignition can be obtained by delaying a few seconds before lighting a modern "coal-effect" household fire.

GASES
Firedamp, or methane, is the most immediate danger in a coalmine, being very flammable. Blackdamp refers to carbon dioxide, which easily accumulates in mines to dangerous proportions and being heavier than air it is found in dips. Whitedamp, or carbon monoxide, is a deadly poison and is a product of incomplete combustion as occurs after a methane - or coal dust explosion. Stinkdamp, or hydrogen sulphide, from sulphur in decay-

ing or burning coal, is extremely poisonous, dissolves in water whence it can be released suddenly, but is easily detected by smell, even in very minute quantities. However, it can be dangerous after an inundation from an old working. Other noxious gases found in coalmines, the products of combustion of coal, diesel fuel or explosives, are sulphur dioxide and the oxides of nitrogen.

GOAF
The "waste", or space left after the coal has been extracted. It may be collapsed or completely stowed with stones and, if not ventilated, it must be walled off. Gas can accumulate in a goaf and a drop in atmospheric pressure can cause it to expand and spill out into the rest of the mine.

HEADING
An advancing place, in the seam.

HEAPSTEAD
The working level platform above ground where the cage is emptied of tubs, allowing gravity to be utilised for screening and loading into waggons.

HEWER
The man who got the coal and filled the tubs. At first most underground workers were hewers but their numbers changed proportionately as mining became more technical. Hewing was considered harder work than most other jobs and was paid accordingly, a shorter shift generally being worked. A hewer used a sharp, short bladed, light pick (q.v.) with a removable blade which was periodically reforged and resharpened in the colliery shops; he carried spare blades and kept others hidden underground near his work. His other equipment consisted of a "cracket" or low stool upon which he sat to swing his pick, and a bundle of numbered metal tokens with attached cord loops which could be pushed through a hole, low in the tub, and fastened to a hook inside, so that his number showed but could not be removed when the tub was full of coal.

At some pits hewers were instructed to work the "whole coal", making the job less dangerous, but usually the coal was undercut to a depth of four feet or so before being wedged or blasted down. A good Hewer was able to swing his pick in any direction to suit the "cleat" of the coal but could expect to produce more when working "bord-ways" (q.v.).

A record output, worthy of mention in a mid 19th Century Newcastle newspaper, was ten tons hewed and filled in a shift by one man at a Sunderland colliery.

Hewers often worked in pairs and, particularly in Northumberland, as "cross-marrows", i.e. in the same place in different, successive shifts, which meant that the hard, slow work of under-cutting could be averaged for pay with the following stage of bringing down the main body of the seam. The coal-cutting machine of the first half of the 20th Century was no more than a mechanical hewer undercutting the coal. Hewing finally died sometime in the 1950's, having lived on in rare pits for obscure reasons. Hand filling on a longwall face, after the coal had been cut and shot down, was regarded as a much softer job, although it would probably overwhelm the modern day manual worker; a filler did not need to go jogging of an evening to keep fit. But it was hewing which epitomised hard pit-work, as Thomas Burt, the first miners' MP, recalled when he wrote in his autobiography, of having had to "face the wall" in his younger days. Certainly it must have been a daunting prospect: the bright, hard wall of coal and only a pick and bare hands with which to provide for a family. The first few inches in were easier where the effects of the last shots had made a crack or two, but after that a badly struck blow might only produce red dust; thousands of jarring, monotonous, aimed swings of "the Queen of Tools" was needed. Mine managers of the day made much of the athletic prowess of their crack hewers, commenting on their daily consumption of meat, allegedly 1¼lbs per day, although others disputed this, assaying only ¼lb; their fine physique was noted, at least as regarding their upper torsos, and in a Paper on the feeding of pit ponies, in 1882, they were favourably contrasted with oat-fed Scottish labourers, an analogy being drawn to show the worth of good feeding for horses. Hewers may have had barrel chests but they usually had "bandy" legs; they were adept at sitting on their "honkers" (q.v.) being used to working while crouched or sitting in low cramped corners.

HOLE
The entrance to the shaft at ground level, used for loading bulky materials, horses etc.

HONKERS
Sitting on one's honkers (haunches). Rarely seen nowadays, but once universal in mining districts. Even on the surface at a week-end, men would sit on their heels in the streets and talk to each other. The custom must have originated in the cramped, seatless conditions of the pits, but children copied and became expert and could sit around the family hearth in this manner, should there be no stool for them.

HORSE ROAD
A roadway at the shaft bottom, driven near-level with, perhaps, a very slight gradient in favour of the load, of sufficient dimensions to permit two-directional traffic of long sets of rollies hauled by large horses. Where possible, such a road would be in the seam, but usually haulage and drainage demands were paramount, and intensive labour with hammer, drill and gunpowder, was necessary to drive it. Such a horse road could be seen at Burradon until it closed in 1975, the triangular shaped sockets of the "old men's" blasting holes still visible in the hard sandstone.

HUMIDITY
The amount of moisture in the air underground became of greater importance as workings went deeper. As stated in a note to the narrative, the rock temperature at 1,200 ft would be 70°F. Allowing for other underground heating effects just about being balanced by the poor, single-shaft, furnace-produced air-current, the temperature would seem to be very tolerable, even pleasant, However, if the air became near saturated, as it well might, if to the natural mine water were added the exhaust steam from pumping or haulage engines, then inconvenience would be felt, and only light clothing tolerated. Damp air cannot cool by evaporation, and sweat

runs in streams down mens' backs. At depths below that of Killingworth the breathing of a hot, saturated atmosphere would have impaired the working capacity of the men, in mines ventilated by the poor furnaces of the early 19th Century. A wet-bulb temperature of 85°F is now regarded as the limit at which any useful work can be done, with saturated air at 80°F to be taken as a more practical upper level. Dry conditions, of course, allow working at much higher temperatures. Also we must remember that, to the miners of those days, work meant hard physical labour: they did not get paid for just going inbye.

High humidity also had the effect of promoting fungal growth on timber, which would seem to be turning into cottonwool, with great white protuberances sapping its strength for supporting the roof.

INBYE
Meaning both the furthest extremity available in a mine and in that general direction; thus the converse OUTBYE means either the shaft bottom or towards it.

KEEKER
Foreman in charge of the heapstead, superintending tipping and screening of coal. The name comes from the Scots "to keek", or to overlook, which is what the Keeker did literally, often having a platform or window where he stood to see who was slacking. He was much feared by the screeners. His official title for legislative purposes was "Heapkeeper".

KEPS
Spring-loaded catches to hold the cage at a shaft bottom or top. Keps are not allowed at intermediate positions for obvious reasons. While the cage is resting on the keps the rope loses its tension but the cage is held steady for the entry or exit of tubs. The keps are released by a lever, held by the banksman, after the winderman has tensioned the rope. Once, in the 1940's at Burradon, the keps jammed, the rope began to pile up on top of the cage and, despite the shouts of the men in the cage, the banksman heaved at the lever until it freed. The cage then plummeted the length of the coils and several men inside were injured.

KIBBLE
A large bucket used to replace a cage especially in a sinking shaft. A modern shaft being sunk uses kibbles running between guide ropes.

KIRVING
Undercutting the coal prior to wedging or blasting down.

KIST
The kist or chest was a large wooden box with a lid and lock. It held the spare explosives, tools of the Deputy, and perhaps those of specialist workmen such as timber drawers, whose sharp axes had to be protected from illicit use by stonemen, who would quickly blunt their razor-sharp edges digging holes in the floor. The kist became the meeting-station for men entering a Deputy's district and nobody was supposed to pass it without the Deputy's knowledge; it was also a convenient seat, and disaffected workmen, upon being asked the whereabouts of the Deputy, might use the stock joke and reply: "Feel the kist. If it's warm he's just left and if it's cold he won't be long in being back." However, in the 19th Century, when the Deputy did the timbering, he had little time for sitting on the kist!

LAID IN
The, hopefully, temporary, partial or complete closure of a mine for economic reasons.

LAID OUT
A tub "Laid-Out" was confiscated by the colliery proprietors without payment to the hewer, because it contained, in the early days, too much small coal and, later, too much stone. The stone was measured to overfill a wooden box and then laid on to flat-sheets, awaiting the surfacing (and dismay) of the hewer. Another reason for confiscation, or even fines at some collieries, was an underweight tub. At first, all such decisions were arbitrarily made by the owner, but later by agreement with the checkweighman according to the agreed rules. The practice always caused anger and slowed the work underground, where filling with graips, careful

searching in a poor light, and tub-shaking to consolidate the load, might be the order of the day. Before tubs, in 1825 a Methodist tract, "A Voice from the Coal Mines," complained of the following imposition: 1 quart of stones per corf - the corf forfeit. 2 quarts of stones per corf - the corf forfeit + 6d. fine. Over this amount - the corf forfeit + 5s.0d. fine.

LANDING
A length of roadway usually level and wide enough for two sets of way, for fulls and empties to be respectively collected and dispersed. Care would be taken in grading the landing so that empty tubs arriving by rope-haulage would pass over a "brow" and their couplings thus lose their tension, allowing them to be disconnected.

LIMBERS
Pronounced 'LIMMERS'. Two wooden shafts connected by an iron bow to a hinged hook for connection to a tub. The limbers were permanently fastened to the pony during its working shift.

LOOSE
Pronounced LOWSE. The end of the shift.

MARRA
Workmate or "marrow", meaning match, doing the same job, sharing the same paynote.

MELL
A corruption of "maul"; a large hammer used for breaking stones or setting heavy timbers. They came in various weights for different trades, a long-handled version being available for withdrawing roof-supports. For really heavy jobs, such as tub-straightening after a collision, word would be sent to the shaft bottom, and a 14 pound job christened "Monday" and normally kept there, would be sent inbye by tub. (It was called Monday because in jest, that was the only day the men were strong enough to use it!)

NEUK
Nook or corner. The neuks on a longwall face are the working places at

each end, situated on the opposite sides of the access road or "gate" to the main length of the face. Hand filling on to a face conveyor belt, a 1950's filler would have to shovel the coal from the innermost extreme twice. This was called "casting". The purpose of the neuk was to provide double packing space for the stones from the gate "caunching" (q.v.); it was also supposed to distribute the overburden weight more evenly over the arch-girders of the gateway, and prevent their distortion. On a low seam the neuk had to be disproportionably longer, to provide sufficient storage for the greater quantity of stone produced, and then had to be ventilated by a "cundy" or conduit left in the pack which, of course, made the neuk longer still. The coal had to be "cast" out and in the next shift the stones had to be "cast" in. A large caunch shot down ready for casting was a daunting sight to the stonemen. During the 1950's the process was mechanised, a steel-rope "slusher", powered by a small winch, dragging the stones into the pack.

NICKING
Vertical cut at the side of a hewer's bord, made to give a clean cut side to the subsequent roadway. By reducing the need for explosives it also resulted in more large coal. Nicking was not popular with Hewers and it was usual to pay extra for it. A strike at Cramlington was over non-payment for nicking.

ONSETTER
The men responsible for getting the full tubs into the cage ready for winding. He would have at least half a dozen lads to help in the immediate vicinity of the shaft bottom, mainly to receive the empties coming down. The onsetter also operated the shaft signals when men were "riding" and controlled their entry to the cage.He earned good money: the shaft was the production bottle neck!

OUTBYE
See inbye.

OVERMAN
General foreman underground in charge of the working of a seam. The

Fore-Overman was senior to the Back-Overman, and kept to the foreshift, usually commencing at 1.00 a.m. or 2.00 a.m., where coal was drawn on two shifts. The backshift would in that case begin at 8.00 a.m. or 9.00 a.m. Ancillary operations would be conducted by the nightshift overman or Master Shifter, who would begin work at 5.00 p.m.

PICK

In the early days the blade was of iron, fitted with tips of steel when that metal was an expensive rarity, but later of cast steel. The usual blade weight was 2lb, but variations of 1lb and 3lb were available for choice. The lighter blades were for "nicking" (q.v.) and for the innermost reach of the "kirving" (q.v.). Hard coal could blunt three or four blades in a shift. The Hewer provided his own pick and drills, but the Owners supplied shovel, mell, wedge and cracket. Sharpening was free but the average hewer had to provide annually for one blade and two shafts. The shaft weighed 2lb. and was fitted with a steel head which could be sprung open to suit any variation of pick eye size. The blade was quickly fastened tightly to the shaft by striking the end against the ground a few times; it was loosened by striking the other end. When William Dryden left Burradon just before the explosion there to work at West Moor, he called at the Burradon pick sharpener's to collect his picks, he testified in court.

PUTTER

Supplier of tubs to the hewers. Until the beginning of the 19th Century haulage was by sledge. Rails were then laid underground (George Stephenson introduced these at Killingworth in 1812) and Putters, either by hand or by pony, took the empties and removed fulls. Putting was the job for young men, usually ex-drivers, before going on to hewing. If he was being served well, a hewer might allow his Putter to fill a tub for himself, or even fill one for him, the Putter's token going through the tub's token-hole, for which he would be credited with the hewing price. For putting tubs, however, he was paid by the score, and would serve two or three hewers throughout the shift, his tokens being tied to the tub handles. He carried a large bundle of corded tokens, or rather his pony did, on the journey inbye. Putters were liable to be bullied by their hewers, who were

anxious to have plenty empties (chumins) available, and were often reduced to tears of frustration when "off the way" in a tight place, sometimes having to walk completely around a pillar of coal to get to the back
of the tub. They sat on the limbers with one hand on the tub handle, and
their heads down; many Putters were killed by not keeping their heads
down.

RAP
Signal given in a shaft or roadway by the pulling of a wire attached to a
hammer which lifted, then on release, "rapped" on an iron sheet. Like
everything else in the pit this required manual effort and the higher numbers of raps were used for the less-frequently required signals; six was for
man-riding, one was always for stop. The pull wires were eventually
replaced by electric bells, powered firstly by batteries and much later by
mains electricity.

ROLLIES
Before tubs and cages were invented, the hazeltwig corves travelled on
rollies to the shaft after their transfer by crane from the putter's flat trams.
A rolley would carry three corves and often ran across the bottom of the
shaft for the attachment of each corve to hooks on the winding rope, one
below the other. The term persists in modern mining with "rolleyway"
meaning the main rope haulage road, and "rolleywayman" the man responsible for its maintenance.

ROPE HAULAGE -ENDLESS
A wire rope running continuously around inbye and outbye return wheels,
driven by a friction pulley powered by a steam engine. Sets of tubs were
attached by grips to the rope.

ROPE HAULAGE-DIRECT
Used on an incline, where the lowering set could draw the rope off the
drive drum by gravity. Hemp ropes were used at first, but were replaced
by chains and then by wire ropes in the 1860's.

ROPE HAULAGE-MAIN AND TAIL

This system, much developed in the 1860's, could work on the level, or cope with rises, falls, and bends, by having two clutched drums and two ropes on the hauler. The main rope pulled the set outbye and at the same time the tail rope was paid out from the hauler, around an inbye return wheel to the back of the set. The tail rope could then be used to pull a new set of empties inbye. A main and tail hauler could pull a set of up to 60 tubs at up to 20 mph. The set was helped around corners by ingenious arrangements of pulleys, rollers and "swan-neck" curves in the track to keep the main rope always central in the roadways. Branch roads could be served by having extra, attachable ropes lying in them, to be coupled into the system when required by a "set-rider" who travelled with the set standing on the back coupling, firmly holding the handles of the last tub (and keeping his head down!).

ROYALTY

Fee paid to the coalowner (q.v.) for his coal, upon its extraction. Royalties were also paid for wayleave, above or below ground, to landowners.

SCORE

Nowadays twenty, but early in the 19th Century, twenty-one tubs in Northumberland, and still twenty-one in Durham until the mid 20th Century.

SHIFTER

One paid by the shift, not on piecework. Generally, older men past their physical prime, or new immigrants to the mining life, not inured by an apprenticeship of putting to the hardship of hewing.

SHOVEL

Early shovels were entirely of wood. To retard wear they would be tipped with copper sheet. Iron shovels, and later steel, were of two sizes: one for coal and one for stone. In the Northern Coalfield they were always of the rounded type. Mid 20th Century, aluminium shovels were introduced, but were quickly withdrawn, along with anything else containing that metal, following the Easington explosion, which showed its incendive sparking propensity.

SPEARS
Rods connected to each other and to the beam of the surface steam engine whose reciprocating motion they followed. They ran through the delivery pipe, operating the piston down the shaft, or, alternatively, ran outside the pipe and operated a force-pump at the shaft-bottom. At New Hartley the dry spears were made of 14 inch square wood reinforced with wrought iron. The wet spears were 12 inch square, in 30 inch pipes.

STAPLE
An underground shaft.

SWALLEY
A dip in an otherwise near-level, in-the-seam roadway where water could collect. Such water had to be cleared by pump, bailing or, where circumstances permitted, by siphon. Where the supply of water was near-infinite, as after an inundation caused by the accidental penetration of neighbouring flooded old workings, such a swalley could be a death-trap to men working inbye of it. These circumstances applied at the Montague Colliery disaster of 1925, taking the lives of most of the 38 men lost, including that of an overman who bravely waded back through the rising water to warn men on the other side of the swalley, six months being required to lower the water level to get to them. At the hideous Heaton flooding of 1815, men so trapped died of starvation.

SYSTEMATIC TIMBERING
Supports set by the workmen, according to a pre-arranged plan, with fixed distances between props. The system was designed by the colliery manager; later on it had to be approved by the Mines Inspector. Additional props may have had to be set for any particular bad ground but an apparently good roof did not mean that any support could be missed out. The method, a great saver of lives, also had the beneficial side-effect of freeing the deputy for other duties, although well into the 20th Century, here and there,

custom demanded that deputies at least saw the props to the right length for their men.

TEN

The "TENTALE" rent paid to the owner of the coal was based on a set of 22 waggons, at first of 19 "bolls" of 2.23 cwt (the most a man could drag in a basket) i.e. 418 bolls, according to Buddle writing of 1805 but by 1840 the Liddell correspondence indicates a ten of 440 bolls. Presumably by then the Killingworth waggons matched those elsewhere in the Great Northern Coalfield and held 20 bolls each.

By the late 19th Century a ten was generally taken to be 8 chaldrons but the 16th Century origin of the word was the ten of the smaller chaldrons of that time to load a Keelboat on the Tyne, the chaldron being the name of the waggon, so called for its shape. Coal was at that time sold by volume.

TIMBERING

The setting of roof supports. Wood props were used, never set direct to the roof but always under a "lid", or "plank" (a prop pre-sawn into two halves lengthwise) or, for more permanent jobs, under a heavy wooden baulk. The composite support was then referred to as "a pair of gears". Timber standing in damp, warm return airways was prone to fungal attack, and spectacular, brilliant-white growths would burst out of the props; roof pressure could snap the planks, and "creep" could bury the bottom of the props, lowering the travelling height of the roadway, until caunching became necessary.

TOKEN

A numbered metal disc with a cord loop attached. Hewers and putters had tokens to fasten to their tubs for piecework payment. The hewer's token was always buried beneath the coal for security, fastened to a hook and usually pulled through a small hole so that its number could be taken before the coal was tipped.

TRAM

A tub base but no container above, except, perhaps, a light frame for holding props or other supplies.

Also, at one time, a term used for two lads doing together one putter's work.

TRAPPER

A very young new-starter's job, opening and closing a wooden ventilation door to allow traffic to pass. With the splitting of air-currents, introduced by Mr. Buddle of Wallsend in the early 19th Century, the trapper's job gradually became redundant, as there were many fewer doors across the coal routes.

TUB

A wooden or wrought-iron box holding from 6cwt, resting on and fastened to a wooden undercarriage with iron wheels and axles, to run on rails. A draw-bar ran underneath from hook to eye so that tubs could be connected in sets of forty or so for rope haulage. A tub was fitted with handles for safer pushing (no nipping of fingers between top of tub and roof!) and with a yoke-hole half way up the box, to take the collar of the limbers for pony haulage.

A tub was so-called from the early system devised to replace corves, where iron tubs were filled and transported underground on bogies, before being lifted off by crane and hoisted singly up the shaft.

The term tub came to mean a full tub. An empty tub was called a "chummin" or "teum one", "teum" or "Toom" being old English for empty.

VIEWER

Original name for Colliery Manager, although a Head Viewer might overlook a group of mines. An Underviewer was an undermanager. Viewing often ran in families, from father to son. After 1872 a manager had to be certificated.

YULEDO

Customary annual Christmas gift from hewer to putter, varying from a currant bun baked by the hewer's wife, to a tub or two filled for him, or hard cash. With the coming of more prosperous days under the National Coal Board, where putting persisted this sentimental emphasis changed to a more worldly one: putters became older and more greedy; they would run inbye and fill an easy tub, taking the best, and could ration their services between hewers or fillers as they felt suited or were best rewarded.

FOOTNOTES

(1) Occasionally the mineral rights had been sold separately from the land and in that case royalties were paid to the new coal owner. Wayleave rights might still have to be paid to the original landowner.

(2) This correspondence is collected in a copybook now in the possesion of Mr. Peter Martin, collector and knowledgeable enthusiast of mining history.

(3) 1 "ten" of 440 "bolls", each of $2.2 \frac{1}{33}$ cwt.

The Strife went on! Soup Kitchen 1926

Bibliography

Kelly's Directory 1900's
Ward's Directory 1851, 1853, 1861
Whellan's Directory 1854, 1863/64, 1867
Pigots Directory 1822 and 1834
Census: 1841 to 1891
Births, Marriages & Burials St. Johns Church, Killingworth
Births, Marriages & Burials Longbenton Parish Church
The Newcastle Journal, (19th Century editions)
The Newcastle Weekly Chronicle, (19th Century editions)
The Newcastle Courant, (19th Century editions)
Transactions of North of England Institute of Mining Engineers
"Lives of the Engineers: George & Robert Stephenson." by Samual Smiles
"Mines and Miners." by L. Simenin
"Annals of Coal Mining and the Coal Trade." by R.L. Galloway
Report on the Employment of Children: Leifchild
Reports to the Commissioners: Tremenhere
"The Town That Was Murdered." by Ellen Wilkinson MP.
Collections held at Northumberland Record Office pertaining to Messrs.
    Buddle, Dunn, Forster, Watson, Wood.
Reports to the Earl of Carlisle from Mathew Liddell of Benton Grange
(held by P. Martin)
Writings of Thomas Burt
H.M. Mining Inspectors' Reports
The Mining Journal
The Colliery Guardian
Report of Northumberland Mines Association, 1944
Reid Report, 1944
Illustrated London News, 1860
"Men of Mark Twixt Tyne & Wear" 1895
Mine Plans of Killingworth, Burradon, Weetslade and Gosforth Collieries,
    now held by the Coal Authority, Burton-on-Trent
"George Stephenson" by W.O. Skeat
Newcastle upon Tyne Accounts and Rent Roll, 1834-36

"The Nature and Origin of Coal and Coal Seams" by A.Raistrick and C.E. Marshall.

"Colliery Working and Management" by H.F. Bulman and R.A.S. Redmayne

The collection of mining textbooks and other records held by the North of England Institute of Mining and Mechanical Engineers at Neville Hall, Newcastle upon Tyne.